METHODS IN MOLECULAR BIOLOGY

Series Editor
John M. Walker
School of Life and Medical Sciences
University of Hertfordshire
Hatfield, Hertfordshire, UK

For further volumes:
http://www.springer.com/series/7651

For over 35 years, biological scientists have come to rely on the research protocols and methodologies in the critically acclaimed *Methods in Molecular Biology* series. The series was the first to introduce the step-by-step protocols approach that has become the standard in all biomedical protocol publishing. Each protocol is provided in readily-reproducible step-by-step fashion, opening with an introductory overview, a list of the materials and reagents needed to complete the experiment, and followed by a detailed procedure that is supported with a helpful notes section offering tips and tricks of the trade as well as troubleshooting advice. These hallmark features were introduced by series editor Dr. John Walker and constitute the key ingredient in each and every volume of the *Methods in Molecular Biology* series. Tested and trusted, comprehensive and reliable, all protocols from the series are indexed in PubMed.

West Nile Virus

Methods and Protocols

Edited by

Fengwei Bai

Department of Cell and Molecular Biology, Center for Molecular and Cellular Biosciences,
The University of Southern Mississippi, Hattiesburg, MS, USA

 Humana Press

Editor
Fengwei Bai
Department of Cell and Molecular
Biology
Center for Molecular and Cellular Biosciences
The University of Southern Mississippi
Hattiesburg, MS, USA

ISSN 1064-3745   ISSN 1940-6029   (electronic)
Methods in Molecular Biology
ISBN 978-1-0716-2762-4   ISBN 978-1-0716-2760-0   (eBook)
https://doi.org/10.1007/978-1-0716-2760-0

Cover Illustration Caption: "Culex quinquefasciatus, the most common mosquito vector of West Nile virus in the United States. Photographed by Donald A. Yee, The University of Southern Mississippi."

This Humana imprint is published by the registered company Springer Science+Business Media, LLC, part of Springer Nature.
The registered company address is: 1 New York Plaza, New York, NY 10004, U.S.A.

Preface

West Nile virus (WNV), a mosquito-borne flavivirus, has caused significant worldwide health concern and is the most common causative viral agent of encephalitis and meningitis in North America. WNV is transmitted between mosquitoes and birds, and humans and other animals can also be infected through infected mosquito bites. It infects humans of all ages, but the elderly and immunocompromised individuals are particularly at risk of developing severe symptoms and even death. Despite intensive laboratory and clinical research of more than two decades, there are still no approved vaccines or antivirals available for human use. The current book provides researchers with the most updated understanding of the basics of WNV biology, virology, epidemiology, and pathogenesis, as well as step-by-step molecular, cellular, and statistical methods to study WNV infection in cell culture, mosquitos, animal models, and human clinical specimen.

Hattiesburg, MS, USA *Fengwei Bai*

Contents

Contributors

ROBERTO AZAR • *Central Virology Laboratory of Israel Ministry of Health, Sheba Medical Center, Ramat Gan, Israel*

FENGWEI BAI • *Department of Cell and Molecular Biology, Center for Molecular and Cellular Biosciences, The University of Southern Mississippi, Hattiesburg, MS, USA*

MEGAN E. CAHILL • *Department of Chronic Disease Epidemiology and the Center for Perinatal, Pediatric and Environmental Epidemiology, Yale School of Public Health, New Haven, CT, USA*

QIANG CHEN • *The Biodesign Institute and School of Life Sciences, Arizona State University, Tempe, AZ, USA*

NICHOLAS B. DEFELICE • *Department of Environmental Medicine and Public Health, Icahn School of Medicine at Mount Sinai, New York, NY, USA; Department of Global Health, Icahn School of Medicine at Mount Sinai, New York, NY, USA*

CHENG-LIN DENG • *Key Laboratory of Special Pathogens and Biosafety, Wuhan Institute of Virology, Center for Biosafety Mega-Science, Chinese Academy of Sciences, Wuhan, China*

LAUREL DUTY • *Department of Cell and Molecular Biology, University of Southern Mississippi, Hattiesburg, MS, USA*

ORAN ERSTER • *Central Virology Laboratory of Israel Ministry of Health, Sheba Medical Center, Ramat Gan, Israel*

ARY FARAJI • *Salt Lake City Mosquito Abatement District, Salt Lake City, UT, USA*

TINGTING GENG • *Department of Immunology, School of Medicine, UConn Health, Farmington, CT, USA*

FREEDOM M. GREEN • *Baylor College of Medicine, National School of Tropical Medicine, Houston, TX, USA; National School of Tropical Medicine, Feigin Biosafety Level 3 Facility, Department of Pediatrics, Division of Tropical Medicine, Baylor College of Medicine and Texas Children's Hospital, Houston, TX, USA*

AREEJ KABAT • *Central Virology Laboratory of Israel Ministry of Health, Sheba Medical Center, Ramat Gan, Israel*

SHAZEED-UL KARIM • *Department of Cell and Molecular Biology, Center for Molecular and Cellular Biosciences, The University of Southern Mississippi, Hattiesburg, MS, USA*

SHARON KARNIELY • *Kimron Veterinary Institute, Israel Ministry of Agriculture, Bet Dagan, Israel*

JOSH LESIO • *The Biodesign Institute and School of Life Sciences, Arizona State University, Tempe, AZ, USA*

NA LI • *Key Laboratory of Special Pathogens and Biosafety, Wuhan Institute of Virology, Center for Biosafety Mega-Science, Chinese Academy of Sciences, Wuhan, China*

HUANLE LUO • *School of Public Health (Shenzhen), Sun Yat-sen University, Shenzhen, China; Key Laboratory of Tropical Disease Control, Sun Yat-sen University, Ministry of Education, Guangzhou, China*

RUTH R. MONTGOMERY • *Department of Internal Medicine, Yale School of Medicine, New Haven, CT, USA*

KRISTY O. MURRAY • *Baylor College of Medicine, National School of Tropical Medicine, Houston, TX, USA*

FARZANA NAZNEEN • *Department of Cell and Molecular Biology, Center for Molecular and Cellular Biosciences, The University of Southern Mississippi, Hattiesburg, MS, USA*

GIRISH NEELAKANTA • *Department of Biomedical and Diagnostic Sciences, College of Veterinary Medicine, University of Tennessee, Knoxville, TN, USA*

BISWAS NEUPANE • *Department of Cell and Molecular Biology, Center for Molecular and Cellular Biosciences, The University of Southern Mississippi, Hattiesburg, MS, USA*

HEPHZIBAH NWANOSIKE • *Baylor College of Medicine, National School of Tropical Medicine, Houston, TX, USA*

AMBER M. PAUL • *Department of Cell and Molecular Biology, University of Southern Mississippi, Hattiesburg, MS, USA; Space Biosciences Division, NASA Ames Research Center, Moffett Field, CA, USA; Blue Marble Space Institute of Science, Seattle, WA, USA; Department of Human Factors and Behavioral Neurobiology, Embry-Riddle Aeronautical University, College of Arts & Sciences (COAS), Daytona Beach, FL, USA*

STEVEN T. PEPER • *Anastasia Mosquito Control District, St. Augustine, FL, USA*

ILIA ROCHLIN • *Center for Vector Biology, Rutgers University, New Brunswick, NJ, USA; Department of Microbiology and Immunology, Center for Infectious Diseases, Stony Brook University, Stony Brook, NY, USA*

SHANNON E. RONCA • *Baylor College of Medicine, National School of Tropical Medicine, Houston, TX, USA; National School of Tropical Medicine, Feigin Biosafety Level 3 Facility, Department of Pediatrics, Division of Tropical Medicine, Baylor College of Medicine and Texas Children's Hospital, Houston, TX, USA*

GILI SCHVARTZ • *Kimron Veterinary Institute, Israel Ministry of Agriculture, Bet Dagan, Israel; Koret School of Veterinary Medicine, The Robert H. Smith, Faculty of Agriculture, Food and Environment, The Hebrew University of Jerusalem, Rehovot, Israel*

MEYTAR SOREK-HAMER • *Environmental Analytics Group (USRA), NASA Ames Research Center, Moffett Field, CA, USA*

AMIR STEINMAN • *Koret School of Veterinary Medicine, The Robert H. Smith, Faculty of Agriculture, Food and Environment, The Hebrew University of Jerusalem, Rehovot, Israel; Veterinary Teaching Hospital, Koret School of Veterinary Medicine, The Robert H. Smith Faculty of Agriculture, Food and Environment, The Hebrew University of Jerusalem, Rehovot, Israel*

HAMEEDA SULTANA • *Department of Biomedical and Diagnostic Sciences, College of Veterinary Medicine, University of Tennessee, Knoxville, TN, USA*

HAIYAN SUN • *The Biodesign Institute and School of Life Sciences, Arizona State University, Tempe, AZ, USA*

KRISHNA KARTHIK VEMURI • *Department of Environmental Medicine and Public Health, Icahn School of Medicine at Mount Sinai, New York, NY, USA*

DOUGLAS E. VETTER • *Department of Otolaryngology – Head and Neck Surgery, University of Mississippi Medical Center, Jackson, MS, USA*

PARMINDER J. S. VIG • *Department of Neurology, University of Mississippi Medical Center, Jackson, MS, USA*

PENGHUA WANG • *Department of Immunology, School of Medicine, UConn Health, Farmington, CT, USA*

TIAN WANG • *Department of Microbiology & Immunology, University of Texas Medical Branch, Galveston, TX, USA; Department of Pathology, University of Texas Medical Branch, Galveston, TX, USA; Institute for Human Infections and Immunity, University of Texas Medical Branch, Galveston, TX, USA*

MATTHEW J. WARD • *Department of Environmental Medicine and Public Health, Icahn School of Medicine at Mount Sinai, New York, NY, USA*

JILL E. WEATHERHEAD • *Baylor College of Medicine, National School of Tropical Medicine, Houston, TX, USA*

HAN-QING YE • *Key Laboratory of Special Pathogens and Biosafety, Wuhan Institute of Virology, Center for Biosafety Mega-Science, Chinese Academy of Sciences, Wuhan, China*

DONALD A. YEE • *School of Biological, Environmental, & Earth Sciences, University of Southern Mississippi, Hattiesburg, MS, USA*

KATHLEEN T. YEE • *Department of Otolaryngology – Head and Neck Surgery, University of Mississippi Medical Center, Jackson, MS, USA*

BO ZHANG • *Key Laboratory of Special Pathogens and Biosafety, Wuhan Institute of Virology, Center for Biosafety Mega-Science, Chinese Academy of Sciences, Wuhan, China*

RONG ZHANG • *Key Laboratory of Medical Molecular Virology (MOE/NHC/CAMS), Shanghai Institute of Infectious Disease and Biosecurity, School of Basic Medical Sciences, Shanghai Medical College, Fudan University, Shanghai, China*

Chapter 1

Introduction to West Nile Virus

Shazeed-Ul Karim and Fengwei Bai

Abstract

West Nile virus (WNV) is a mosquito-borne, single-stranded, positive-sense RNA virus belonging to the *Flaviviridae* family. After WNV gains entry through an infected mosquito bite, it replicates in a variety of human cell types and produces a viremia. Although the majority of infected individuals remain asymptomatic, the manifested symptoms in some people range from a mild fever to severe neurological disorder with high morbidity and mortality. In addition, many who recover from WNV neuroinvasive infection present with long-term deficits, including weakness, fatigue, and cognitive problems. Since entering the USA in 1999, WNV has become the most common mosquito-borne virus in North America. Despite the intensive research over 20 years, there are still no approved vaccines or specific treatments for humans, and it remains an urgent need to understand the pathogenesis of WNV and develop specific therapeutics and vaccines.

Key words West Nile virus, Viral proteins, Transmission, Pathogenesis

1 Viral Genome and Proteins

WNV particle is spherical in shape with approximately 50 nm in diameter. It has an icosahedral nucleocapsid encircled by a lipid envelope [1]. The virus has an approximately 11 kb-long, single-stranded, positive-sense RNA genome containing a single open reading frame (ORF) flanked by untranslated regions (UTRs) at both ends, which extend to 96 nucleotides at the 5′ end and 631 nucleotides at the 3′ end in the epidemic strain WNV NY99 [2]. Like other flaviviruses, the WNV genome is 5′ capped with no polyadenylation tail at the 3′ end. The UTRs of the viral RNA genome assist in replication, transcription, translation, and packaging [3–5]. The viral RNA is translated as a polyprotein, which is cleaved by cellular and viral proteases into three structural proteins (capsid, envelope, and pre-membrane) and seven nonstructural proteins (NS1, NS2A, NS2B, NS3, NS4A, NS4B, and NS5) [2].

Fengwei Bai (ed.), *West Nile Virus: Methods and Protocols*,
Methods in Molecular Biology, vol. 2585, https://doi.org/10.1007/978-1-0716-2760-0_1,

1.1 Structural Proteins

WNV capsid (C) protein encapsulates viral RNA to construct a nucleocapsid. The N- and C-terminals of the C protein have been reported to aid in RNA folding during viral replication, while the middle section of the C protein has been reported to involve direct interaction with viral RNA [6]. The envelope (E) protein is a transmembrane protein, and its ectodomain contains three structural domains (DI, DII, and DIII). DI is comprised of a single glycosylation motif, DII contains an internal fusion loop, and DIII is an immunoglobulin (Ig)-like domain, which plays a crucial role in binding to cellular receptors and entry of the virus into cells [6, 7]. The pre-membrane/membrane (prM/M) protein is a small, glycosylated protein, which is essential in the formation and egression of progeny WNV particles. PrM protein, along with E protein, facilitates the formation of immature virions and inhibits a premature fusion during virus egress. Viral particles with immature prM are noninfectious until prM cleaves to M to stimulate the membrane-fusion activities of the virus [7].

1.2 Nonstructural Proteins

NS1 is a glycoprotein, which is found intracellularly, on the infected cell surface, or secreted by infected cells in the blood. The intracellular NS1 mediates viral replication and assembly, whereas the secreted NS1 regulates viral immune responses in the host [7, 8]. NS1 can bind to the regulatory protein factor H inhibiting the complement activation. It also interacts with C4, downregulates the production of type I interferon (IFN), and deactivates the classical and lectin pathways of the complement system. NS1 of WNV has also been reported to stimulate hyperpermeability of human brain endothelial cells to cause neurological disorders and facilitate encephalitis [9].

Both NS2A and NS2B are small transmembrane proteins. NS2A is a hydrophobic protein associated with the membrane of the endoplasmic reticulum (ER). It has been reported to form the scaffold for the viral replication complex and plays an important role in replicating the WNV genome, inhibiting host immune responses, producing virus-induced membrane structures, and assembling virion [10–12]. NS2B has been suggested to couple with NS3 protease cleaving viral protein to facilitate NS3 proteolytic activity and contribute to the host cell apoptosis and neuropathogenesis [7, 13].

NS3 is the best-characterized multifunctional protein, which is highly conserved in all flaviviruses. NS3 couples with the cofactor NS2B to form the NS2B-NS3 protease that cleaves the viral polyprotein into the structural and nonstructural proteins [14]. NS3 is also involved in helicase, nucleoside triphosphatase, and RNA triphosphatase activities that are essential to viral genome replication. NS3 helicase segregates double-stranded (ds) RNA during viral replication and promotes WNV resistance to antiviral action of 2′,

5′–oligoadenylate synthetase 1b (Oas1b). In addition, the WNV NS3 helicase has been shown to inhibit type I IFN-mediated antiviral responses [7].

NS4A and NS4B are small hydrophobic proteins connected by a 2K peptide composed of 23 amino acid residues. Both proteins are associated with membrane rearrangement in infected host cells and inhibiting IFN responses during WNV infection [7]. NS4A and NS4B have been shown to develop the scaffold for viral genome replication, inhibition of host immune responses, production of virus-induced membrane structures, and virion assembly [10–12].

NS5 is the largest nonstructural protein encoded by the viral genome. It is composed of N-terminal methyltransferase (MTase) activity and C-terminal RNA-dependent RNA-polymerase (RdRp) activity. The NS5 MTase activity is responsible for viral genome 5′ capping, which mimics eukaryotic mRNA allowing the use of the host protein synthesis components for viral gene translation [15–17]. Like many other RNA viruses, WNV NS5 RdRp lacks a proofreading mechanism; thus, the viral populations display a variable level of sequence diversity that may facilitate the viral evolution in response to selective pressures [7]. Besides the functions in viral genome replication, NS5 has also been suggested to have a significant role in the viral pathogenesis by inhibiting type I IFN responses of the host [18].

In summary, the structural proteins assist in forming virion structure, attachment, and entry, whereas the nonstructural proteins are responsible for viral genome replication, polyprotein translation, virion assembly, maturation, egress, and regulation of host immune responses. The primary functions of the WNV proteins are summarized in Table 1.

2 Epidemiology

WNV was first isolated from a female patient in 1937 in Uganda [19]; however, the first indication that its infection led to meningitis and encephalitis in humans was in the outbreaks in Israel in 1957 and France in the early 1960s [19–21]. WNV came into the spotlight when several outbreaks were reported in some parts of North Africa, Europe, and Israel, with severe neurological diseases cases and death incidence in the 1990s [22–24]. In 1999, WNV gained its entry into North America in New York City, and within 3 years, it had spread rapidly to most of the parts of the US and the neighboring countries, including Canada, Mexico, the Caribbean, and Central America [20, 21, 25]. In 2012, the USA witnessed an epidemic of WNV with 5,674 confirmed cases, among which 286 people died and over 50% developed neuroinvasive diseases [26–28]. Since 1999, approximately 25,000 neurological

Table 1
Functions of WNV proteins

Proteins		Functions
Structural proteins	Capsid (C)	RNA folding, nucleocapsid construction, and regulation of cell apoptosis
	Envelope (E)	Receptor attachment and entry
	Pre-membrane/membrane (prM/M)	Formation and maturation of viral particle
Nonstructural proteins	NS1	Viral replication and assembly, inhibition of the complement activation, and type I interferon production
	NS2A	Scaffold formation for viral replication, inhibition of host immune response, and virion assembly
	NS2B	Cofactor of NS2B-NS3 protease and polyprotein cleavage
	NS3	Proteolytic activity, helicase, IFN inhibition, and inhibition of antiviral response
	NS4A	Cofactor of NS3 helicase, membrane rearrangement, IFN inhibition, and regulation of immune response
	NS4B	Membrane rearrangement, IFN inhibition, and regulation of immune response
	NS5	Methyltransferase, RNA polymerase, IFN inhibition, and regulation of immune response

complications and over 2,300 deaths associated with WNV infections have been recorded in the USA [21, 29]. In 2018, a massive outbreak of WNV happened throughout 15 countries in Europe, with around 2,000 confirmed cases [30]. WNV has now spread throughout the rest of the world except Antarctica, and it is now considered the most important causative agent of human viral encephalitis worldwide. Therefore, there is an unmet need to understand the pathogenesis of WNV and develop specific therapeutics and vaccines.

3   Transmission

WNV transmission in nature maintains in a cycle between mosquitos and various bird species. Different species of *Culex* mosquitoes are the main transmission vectors of WNV worldwide. *Culex tarsalis* and *C. pipiens* have been suggested to be the primary vectors in the western and eastern parts of the USA, respectively [31–33]. The American robin (*Turdus migratorius*) is the most important host for the maintenance and transmission of WNV in the USA [34]. After a mosquito acquires WNV from a blood meal, the virus infects the midgut epithelial cells and starts its replication. WNV then travels to the salivary glands of mosquitoes via circulating fluid. When this infected mosquito feeds on humans or other animals, WNV may be inoculated into the host skin. WNV replicates in a variety of cells in humans, such as neutrophils, macrophages, and keratinocytes and generates a viremia. The viral load in blood in an infected mammal usually is not high enough as in birds to be transmitted to another mosquito; therefore, humans and other animals are considered as the dead-end hosts in the WNV transmission cycle [35]. Besides mosquito bites, the transmission of WNV is also possible through blood transfusion, organ transplantation, breastfeeding, and laboratory-acquired infection [32–34].

4 Clinical Symptoms and Pathogenesis of WNV in Humans

WNV infection in humans is predominantly asymptomatic (80%), but it may result in a spectrum of diseases in about 20% of infected people. Among those symptomatic individuals, most may have a fever, malaise, headache, backache, myalgias, arthralgias, gastrointestinal symptoms (nausea, vomiting, or diarrhea), and maculopapular rash. About 1 in 150 people infected with WNV develop severe neuroinvasive diseases due to the viral infection in the central nervous system (CNS). Symptoms of neuroinvasive disease include high fever, severe headache, neck stiffness, confusion, stupor, tremors, seizures, muscle weakness or paralysis, and focal neurological deficits. About 10% of infected people develop severe neuroinvasive diseases such as flaccid paralysis, encephalitis, and meningitis die. Risk factors for encephalitis and death include advanced age, a history of cardiovascular or chronic diseases, and immunosuppression. Among those infected elderlies who survive WNV neuroinvasive diseases, as many as 50% may develop post-illness morbidity, such as fatigue, dizziness, difficulty concentrating, depression, anxiety, sleep disruption, recurrent headaches, and even autoimmune diseases, despite the clearance of infectious viruses from the body within a few weeks.

After WNV enters the human body through a mosquito bite, the virus replicates in keratinocytes and resident skin dendritic cells, i.e., Langerhans cells (LCs). Through blood and lymphoid circulations, WNV travels to and infects the peripheral organ such as the spleen, liver, and kidney. A low-level viremia is generated, which usually lasts for a few days. Infected LCs travel to the lymph nodes and spleen after being activated by WNV antigen, leading to T-cell activation [36]. Following WNV infection, dendritic cells produce a large amount of type I IFNs that may suppress the spread of the virus in the early phase of the infection. However, the signaling is decreased in DCs of the aged individuals infected with WNV, which could partially explain why WNV infection causes severe diseases in aged patients [37, 38]. Although the mechanisms of WNV entry into the CNS are not fully understood, multiple possible pathways have been suggested and verified in animal models. These pathways include flow through the tight junctions of the blood-brain barrier (BBB) from the blood circulation, direct infection of endothelial cells in the cerebral microvasculature, infection of olfactory neurons, "Trojan horse" transport via infected leukocytes, and/or direct axonal retrograde transport from peripheral neurons [9]. In the CNS, WNV can infect various types of cells, including neurons, astrocytes, and microglial cells, resulting in cell apoptosis or necrosis, inflammation of tissue and/or membrane of the brain, which may lead to encephalitis and meningitis.

Acknowledgement

This work was supported in part by the National Institute of Allergy and Infectious Diseases of the National Institutes of Health R15AI135893 (F.B.).

References

1. Mukhopadhyay S, Kim BS, Chipman PR et al (2003) Structure of West Nile virus. Science 302(5643):248. https://doi.org/10.1126/science.1089316

2. Markoff L (2003) 5′- and 3′-noncoding regions in flavivirus RNA. Adv Virus Res 59: 177–228. https://doi.org/10.1016/s0065-3527(03)59006-6

3. Friebe P, Harris E (2010) Interplay of RNA elements in the dengue virus 5′ and 3′ ends required for viral RNA replication. J Virol 84(12):6103–6118. https://doi.org/10.1128/JVI.02042-09

4. Khromykh AA, Meka H, Guyatt KJ et al (2001) Essential role of cyclization sequences in flavivirus RNA replication. J Virol 75(14): 6719–6728. https://doi.org/10.1128/JVI.75.14.6719-6728.2001

5. Shi PY, Brinton MA, Veal JM et al (1996) Evidence for the existence of a pseudoknot structure at the 3′ terminus of the flavivirus genomic RNA. Biochemistry 35(13): 4222–4230. https://doi.org/10.1021/bi952398v

6. Acharya D, Bai F (2016) An overview of current approaches toward the treatment and prevention of West Nile virus infection. Methods Mol Biol 1435:249–291. https://doi.org/10.1007/978-1-4939-3670-0_19

7. Bai F, Thompson EA (2021) West Nile Virus (Flaviviridae). In: Bamford DH, Zuckerman M (eds) Encyclopedia of virology, 4th edn. Academic Press, Oxford, pp 884–890. https://doi.org/10.1016/B978-0-12-809633-8.21504-5

8. Westaway EG, Mackenzie JM, Khromykh AA et al (2002) Replication and gene function in Kunjin virus. Curr Top Microbiol Immunol 267:323–351. https://doi.org/10.1007/978-3-642-59403-8_16

9. Bai F, Town T, Pradhan D et al (2007) Antiviral peptides targeting the West Nile virus envelope protein. J Virol 81(4):2047–2055. https://doi.org/10.1128/JVI.01840-06

10. Grant D, Tan GK, Qing M et al (2011) A single amino acid in nonstructural protein NS4B confers virulence to dengue virus in AG129 mice through enhancement of viral RNA synthesis. J Virol 85(15):7775–7787. https://doi.org/10.1128/JVI.00665-11

11. Wicker JA, Whiteman MC, Beasley DW et al (2006) A single amino acid substitution in the central portion of the West Nile virus NS4B protein confers a highly attenuated phenotype in mice. Virology 349(2):245–253. https://doi.org/10.1016/j.virol.2006.03.007

12. Xie X, Wang QY, Xu HY et al (2011) Inhibition of dengue virus by targeting viral NS4B protein. J Virol 85(21):11183–11195. https://doi.org/10.1128/JVI.05468-11

13. Ramanathan MP, Chambers JA, Pankhong P et al (2006) Host cell killing by the West Nile virus NS2B-NS3 proteolytic complex: NS3 alone is sufficient to recruit caspase-8-based apoptotic pathway. Virology 345(1):56–72. https://doi.org/10.1016/j.virol.2005.08.043

14. Lindenbach BD, Rice CM (2001) Flaviviridae: the viruses and their replication. Fields' Virol, 4th ed., pp. 991–1110

15. Ray D, Shah A, Tilgner M et al (2006) West Nile virus 5'-cap structure is formed by sequential guanine N-7 and ribose 2'-O methylations by nonstructural protein 5. J Virol 80(17): 8362–8370. https://doi.org/10.1128/JVI.00814-06

16. Tan BH, Fu J, Sugrue RJ et al (1996) Recombinant dengue type 1 virus NS5 protein expressed in Escherichia coli exhibits RNA-dependent RNA polymerase activity. Virology 216(2):317–325. https://doi.org/10.1006/viro.1996.0067

17. Henderson BR, Saeedi BJ, Campagnola G et al (2011) Analysis of RNA binding by the dengue virus NS5 RNA capping enzyme. PLoS One 6(10):e25795. https://doi.org/10.1371/journal.pone.0025795

18. Keating JA, Bhattacharya D, Lim PY et al (2013) West Nile virus methyltransferase domain interacts with protein kinase G. Virol J 10:242. https://doi.org/10.1186/1743-422X-10-242

19. Smithburn KC, Hughes TP, Burke AW et al (1940) A neurotropic virus isolated from the blood of a native of Uganda. Am J Trop Med 20:471–472. https://doi.org/10.4269/ajtmh.1940.s1-20.471

20. Bai F, Thompson EA, Vig PJS et al (2019) Current understanding of West Nile virus clinical manifestations, immune responses, Neuroinvasion, and immunotherapeutic implications. Pathogens. https://doi.org/10.3390/pathogens8040193

21. Spigland I, Jasinska-Klingberg W, Hofshi E et al (1958) Clinical and laboratory observations in an outbreak of West Nile fever in Israel in 1957. Harefuah 54(11):275–280. PMID: 13562703

22. Murgue B, Murri S, Triki H et al (2001) West Nile in the Mediterranean basin: 1950-2000. Ann N Y Acad Sci 951:117–126. https://doi.org/10.1111/j.1749-6632.2001.tb02690.x

23. Bin H, Grossman Z, Pokamunski S et al (2001) West Nile fever in Israel 1999-2000: from geese to humans. Ann N Y Acad Sci 951:127–142. https://doi.org/10.1111/j.1749-6632.2001.tb02691.x

24. Tsai TF, Popovici F, Cernescu C et al (1998) West Nile encephalitis epidemic in southeastern Romania. Lancet 352(9130):767–771. https://doi.org/10.1016/s0140-6736(98)03538-7

25. Gubler DJ (2007) The continuing spread of West Nile virus in the western hemisphere. Clin Infect Dis 45(8):1039–1046. https://doi.org/10.1086/521911

26. Mann BR, McMullen AR, Swetnam DM et al (2013) Molecular epidemiology and evolution of West Nile virus in North America. Int J Environ Res Public Health 10(10): 5111–5129. https://doi.org/10.3390/ijerph10105111

27. Centers for Disease Control and Prevention (CDC). West Nile virus disease cases and deaths reported to CDC by year and clinical presentation, 1999–2020. https://www.cdc.gov/westnile/statsmaps/cumMapsData.html

28. West Nile virus and other arboviral diseases-United States, 2012 (2013) MMWR Morb Mortal Wkly Rep 62(25):513–517. PMID: 23803959

29. Chancey C, Grinev A, Volkova E et al (2015) The global ecology and epidemiology of West Nile virus. Biomed Res Int 376230. https://doi.org/10.1155/2015/376230

30. Kaiser JA, Barrett ADT (2019) Twenty years of progress toward West Nile virus vaccine development. Viruses 11(9):823. https://doi.org/10.3390/v11090823

31. Kilpatrick AM, Daszak P, Jones MJ et al (2006) Host heterogeneity dominates West Nile virus transmission. Proc Biol Sci 273(1599): 2327–2333. https://doi.org/10.1098/rspb.2006.3575

32. Reisen WK, Fang Y, Martinez VM et al (2005) Avian host and mosquito (Diptera: Culicidae) vector competence determine the efficiency of West Nile and St. Louis encephalitis virus transmission. J Med Entomol 42(3):367–375. https://doi.org/10.1093/jmedent/42.3.367

33. Kilpatrick AM, Kramer LD, Jones MJ et al (2006) West Nile virus epidemics in North America are driven by shifts in mosquito feeding behavior. PLoS Biol 4(4):e82. https://doi.org/10.1371/journal.pbio.0040082

34. Hamer GL, Kitron UD, Goldberg TL et al (2009) Host selection by Culex pipiens mosquitoes and West Nile virus amplification. Am J Trop Med Hyg 80(2):268–278. PMID: 19190226

35. Bowen RA, Nemeth NM (2007) Experimental infections with West Nile virus. Curr Opin Infect Dis 20(3):293–297. https://doi.org/10.1097/QCO.0b013e32816b5cad

36. Johnston LJ, Halliday GM, King NJ (2000) Langerhans cells migrate to local lymph nodes following cutaneous infection with an arbovirus. J Invest Dermatol 114(3):560–568. https://doi.org/10.1046/j.1523-1747.2000.00904.x

37. Kovats S, Turner S, Simmons A et al (2016) West Nile virus-infected human dendritic cells fail to fully activate invariant natural killer T cells. Clin Exp Immunol 186(2):214–226. https://doi.org/10.1111/cei.12850

38. Qian F, Wang X, Zhang L et al (2011) Impaired interferon signaling in dendritic cells from older donors infected in vitro with West Nile virus. J Infect Dis 203(10):1415–1424. https://doi.org/10.1093/infdis/jir048

Chapter 2

Quantification of West Nile Virus by Plaque-Forming Assay

Biswas Neupane and Fengwei Bai

Abstract

The plaque-forming assay is a gold standard technique to determine the concentration of infectious viral particles. In this assay, lytic viruses infect and lyse the cells but are immobilized due to the presence of an agarose-containing overlay medium. The progeny viruses can only spread locally to and kill the adjacent cells and finally form a clear zone or plaque after staining the live cells. The number of plaques formed can be theoretically considered as the initial number of the infectious viral particles present in the sample and hence can be expressed as plaque-forming units (PFU) in a volume of the sample. Here, we provide a step-by-step method to carry out a plaque-forming assay to determine the titer of West Nile virus in a cell culture medium, which also can be adapted to other lytic viruses of eukaryotic cells.

Key words West Nile virus, Plaque-forming assay

1 Introduction

The first plaque-forming assay method was developed in 1952 to determine titers of infectious lytic animal viruses [1]. Although various modern techniques are being used for viral quantification, such as assays based on quantitative reverse transcription-polymerase chain reaction (qRT-PCR), flow cytometry, and immunostaining, these assays have their limitations. For instance, they cannot quantify the replication of infectious viral particles, and the plaque-forming assay is still considered as the gold standard approach to determine a concentration of a lytic virus [2].

In a plaque-forming assay, a virus-containing sample is serially diluted then applied to a confluent monolayer of cells. Infectious viral particles attach and enter the host cells during a short time of incubation, and then an overlay medium is applied over the cells. This overlay consists of two major components: nutrition for the cells and agarose for the formation of the gel. Due to the immobilization of the agarose gel, the virus replicates inside the initially infected cell, and the progeny viruses can only spread locally and kill the adjacent cells in the monolayer. This cycle of the infection and

Fengwei Bai (ed.), *West Nile Virus: Methods and Protocols*,
Methods in Molecular Biology, vol. 2585, https://doi.org/10.1007/978-1-0716-2760-0_2,

lysis of the neighboring cells leads to the formation of plaques, which are the circular zones of cell death. Theoretically, each infectious particle produces a plaque and eventually becomes large enough to be visible to the naked eyes. West Nile virus (WNV) plaques usually will be visible within 3–4 days of infection. The plaques can be counted under a microscope or counterstained with neutral red or crystal violet to aid observation with the naked eyes. The viral titer can be expressed as plaque-forming units (PFU) per milliliter of the sample. We have used this assay to quantify the titer of WNV [3] or other viruses, i.e., Zika and chikungunya viruses [4, 5].

2   Materials

1. Virus: a WNV sample to be tested.
2. Cells: Vero cells (ATCC CCL-81, *see* **Note 1**).
3. Culture medium: Dulbecco's Modified Eagle Medium (DMEM) with 10% FBS (fetal bovine serum) and 1% P/S (penicillin/streptomycin).
4. Overlay medium: 5% SeaPlaque agarose (SPA) prepared by mixing 5 g of SPA in 100 mL of PBS and autoclaving to ensure sterility.
5. Neutral red: 0.33% solution.
6. Plates: Six-well cell culture plates.
7. CO_2 incubator.
8. Others: Serological pipettes, pipette aid, and pipette tips.

3   Methods

3.1   Cell Plating

1. Plate healthy Vero cells (90% to 100% confluency after overnight incubation) in six-well plates by plating $5–6 \times 10^5$ cells per well in a final volume of 2 mL of culture medium (*see* **Note 2**).
2. Incubate the plates in a 37 °C incubator with 5% CO_2, overnight.

3.2   Dilution and Infection

1. Next day, check the confluency and viability of the cells before starting the assay. The cells should be expressing their standard cellular morphology at confluency between 90% and 100% (*see* **Note 3**).
2. Perform a tenfold serial dilution of the viral stock using the cell culture medium as the diluent. WNV sample can be serially diluted by adding 50 µl of the sample to 450 µl of the culture

medium successively (*see* **Note 4**). Consider including a positive control with known viral titer and a negative control without viruses.

3. Completely remove the medium from the plates, and add 900 µl of prewarmed freshly prepared culture medium to each well (*see* **Note 5**). Add 100 µl of the serially diluted virus samples, positive control, and negative control to the prelabeled wells, and incubate the plate in a 37 °C incubator with 5% CO_2 for 1 h (*see* **Notes 6–8**).

4. During the incubation time, prepare the overlay medium (SPA). Heat 10 mL of 5% SPA in a microwave, add 40 mL of prewarmed (37 °C) culture medium, and gently mix well. Place the overlay solution in a 37 °C water bath until use to equilibrate the temperature (*see* **Notes 9–11**).

5. After incubation, completely remove the virus-containing medium from the wells, and add 2 mL of prewarmed overlay medium. Make sure that the solution is warm but is not too hot to cause cell death (*see* **Note 12**).

6. After the agarose solidifies (within 10–15 min), incubate plates at 37 °C in 5% CO_2 until distinct plaques are formed (*see* **Note 13**). WNV plaques will develop in 3 to 4 days of incubation. Plaque development can be monitored periodically by checking the wells under a microscope (*see* **Note 14**).

3.3 Staining

Once plaques can be observed, they can be counterstained with neutral red. Prepare a working solution by mixing 5 mL of neutral red in 10 mL of PBS, and add 500 µl of diluted neutral red solution to the wells, and then incubate the plates in a 37 °C incubator with 5% CO_2 for 3–4 h.

3.4 Determining Viral Titers

1. Carefully observe the size and morphology of the plaques as in Fig. 1. Use negative control and positive control as references. Count plaques observed in each well, and take the average of the triplicates (*see* **Notes 15** and **16**). Count the plaques if they are within the range of 5 to 100 per well (*see* **Note 17**).

2. Calculate the viral titer by using the following equation (*see* **Note 18**).

$$PFU/mL = \frac{\text{Average number of plaques}}{D \times V}$$

where D = dilution and V = sample volume.

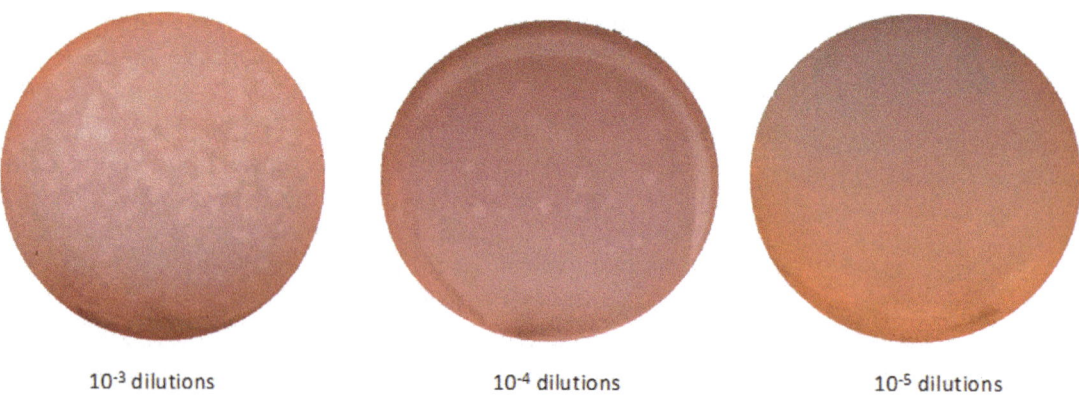

10^{-3} dilutions 10^{-4} dilutions 10^{-5} dilutions

Fig. 1 Representative images of WNV plaques WNV plaques after staining with neutral red solution for 4 h. The lightly stained regions represent the plaques formed after incubation for 4 days.

4   Notes

1. Depending on the experiment, different cells can be used for growing the virus.

2. The viability of the cells should be checked before the experiment. If the cells are not healthy, they may start falling off the plate, and plaques will not form.

3. The confluency of the cells is crucial for the success of the assay. It can be achieved by plating $5–6 \times 10^5$ cells per well in a final volume 2 mL of the culture medium in a six-well plate. Alternatively, 12-well plates can also be used for plaque assay. For this, cells should be plated at a density of 2×10^5 cells per well in a final volume 1 mL of the culture medium. If the cells do not look confluent enough, wait for a few more hours to reach the desired level of confluency.

4. All the procedures with live WNV should be carried out inside a biosafety cabinet in a biosafety level 3 (BSL-3) laboratory.

5. While changing the media, it should be done carefully and slowly to avoid touching the cell monolayer as it may disturb the cells in the region and eventually affect plaque formation.

6. While adding viruses to the wells, the volume of inoculum should be considered. It should be just enough to cover the entire region of the cells. 1 mL is appropriate for a well of six-well plates.

7. Including a positive and negative control to the assay will be helpful for troubleshooting, if needed.

8. The plates should be gently rotated every 20 min during incubation to make the viruses distribute through the wells.

9. Autoclaved 5% SPA can be aliquoted to 10 mL per tube in 50 mL sterilized tubes and stored in a refrigerator.

10. Special attention is required while preparing the overlay medium. Heat 10 mL of 5% SPA in a microwave to melt but not boil. Boiling may result in overflowing of SPA from the tube.

11. Using freshly prepared culture medium to make an overlay medium is recommended. Culture medium should be pre-warmed at 37 °C before adding to 5% SPA. The use of cold culture medium results in solidifying the agarose in the tubes before adding to the wells.

12. While adding to the wells, the overlay medium should be warm (40 to 45 °C) but not too hot to cause cell death.

13. After adding the overlay medium into the wells, allow it to solidify in a running biosafety cabinet for 10–15 min with the lids open. Don't move the plates during this time, for it may interrupt the solidification and damage the cell monolayer.

14. During incubation, it's suggested to monitor the development of the plaques periodically. The wells can be observed under a microscope to ensure that the plaques do not overgrow. When plaques are overgrown, they may fuse and affect the results.

15. While counting the plaques, using a lightbox will aid in better visualization of the plaques. The plate can be placed on a lightbox with the base facing you.

16. For a more accurate counting, it is suggested to use a Sharpie pen to mark the counted plaques.

17. For calculating the viral plaques, it's better to use the dilution in which the plaque numbers lie in the range between 5 and 100.

18. An example, suppose that the numbers of the plaques are 28, 32, and 30 in triplicates at the dilution of 10^{-6}. Suppose the virus sample volume is 100 μl (0.1 mL).

$$\text{PFU/ml} = \frac{\text{Average number of plaques}}{D \times V}$$
$$= \frac{30}{10^{-6} \times 0.1}$$
$$= 3 \times 10^8 \ \text{PFU/ml}$$

Acknowledgments

The authors thank Ms. Farzana Nazneen for providing the WNV plaque images. This work was supported in part by the National Institute of Allergy and Infectious Diseases of the National Institutes of Health R15AI135893 (F.B.).

References

1. Dulbecco R (1952) Production of plaques in the monolayer tissue cultures by single particles of an animal virus. Proc Natl Acad Sci U S A 38(8):747–752

2. Juarez D, Long KC, Aguilar P et al (2013) Assessment of plaque assay methods for alphaviruses. J Virol Methods 187(1):185–189

3. Bai F, Wang T, Pal U et al (2005) Use of RNA interference to prevent lethal murine West Nile virus infection. J Infect Dis 191:1148–1154

4. Neupane B, Fendereski M, Nazneen F et al (2021) Murine trophoblast stem cells and their differentiated cells attenuate Zika virus in vitro by reducing glycosylation of the viral envelope protein. Cells 10(11):3085

5. Neupane B, Acharya D, Nazneen F et al (2020) Interleukin-17A facilitates chikungunya virus infection by inhibiting IFN-α2 expression. Front Immunol 11:588382

Chapter 3

Viral Titer Quantification of West Nile Virus by Immunostaining Plaque Assay

Na Li, Cheng-Lin Deng, Bo Zhang, and Han-Qing Ye

Abstract

Immunostained plaque assay based on the specific antibody binding to viral antigen enables the detection and titration of virus infectivity, especially for viruses that could not form plaques using the classical crystal violet or neutral red staining methods. Here we describe the application of this method to quantify viral titers of wild-type West Nile virus (WNV-WT) and replication-defective WNV-ΔNS1 virus.

Key words Immunostained plaque assay, Viral antigen, Titer quantification, West Nile virus, WNV-ΔNS1 virus

1 Introduction

For most viruses, including West Nile virus (WNV), the classical plaque assay based on crystal violet or neutral red staining is applied for the quantification of viral titers, as each plaque represents an infectious virus particle [1]. Nonetheless, there still exist some exceptions when no or small, fuzzy plaques occur as a result of specific passage history of viruses, mutations or deletions in the genome of mutant viruses [2–5]. Under the circumstances, the employment of immunostaining approach into the plaque assay can overcome the limitations [6]. In immunostained plaque assay, viral antigens expressed by infected cells are immunostained with labeled antibodies, and the foci are visualized by chromogenic reactions [1, 7]. This method shortens the detection time and extends the storage time of samples in comparison with the classical plaque staining approach [8].

Previously, we reported that WNV-ΔNS1 virus (with the deletions of residues 4–298) could replicate in BHK$_{NS1}$ cell line (a BHK-21 cell line stably expressing WNV NS1 protein) efficiently but were not replicable in naïve BHK cells [9, 10]. However, this replication-defective WNV-ΔNS1 could not produce any clear

Fengwei Bai (ed.), *West Nile Virus: Methods and Protocols*,
Methods in Molecular Biology, vol. 2585, https://doi.org/10.1007/978-1-0716-2760-0_3,

plaques using the conventional crystal violet staining approach in BHK$_{NS1}$ cells. Therefore, an immunostaining method, namely, immunostained plaque assay, was performed to quantify the viral titer of the WNV-ΔNS1 virus. In this chapter, we describe the detailed information about this assay using wild type WNV (WNV-WT) and WNV-ΔNS1 viruses.

2 Materials

2.1 Cells, Viruses, and Antibodies

1. BHK-21 cells (ATCC® CCL-10™).
2. BHK$_{NS1}$ cells. The stable BHK-21 cell line expressing WNV NS1 protein (named BHK$_{NS1}$) is established as described previously [9, 10].
3. WNV-WT viruses. Recombinant WNV-WT viruses are generated by electroporation of BHK-21 cells with in vitro transcribed viral genomic RNA from the linearized infectious cDNA clone of strain 3356 from New York City [11, 12]. The supernatants of transfected cells are harvested at 72 h post transfection and frozen aliquots at −80 °C.
4. WNV-ΔNS1 viruses. The recombinant replication-defective WNV-ΔNS1 viruses are obtained by transfection of BHK$_{NS1}$ cells with the transcribed viral genomic RNA from the linearized infectious cDNA clone of WNV-ΔNS1 [9, 10]. The supernatants of transfected cells are harvested at 72 h post transfection and frozen aliquots at −80 °C.
5. 4G2 antibody: The primary monoclonal antibody 4G2 is against the flavivirus envelope protein.
6. The secondary horseradish peroxidase (HRP)-conjugated goat anti-mouse IgG antibody.

2.2 Cell Culture

1. 10% DMEM: DMEM medium containing 10% heat-inactivated FBS (fetal bovine serum) and 1% penicillin-streptomycin solution. Store at 4 °C.
2. BHK-21 cell culture: Grow in 10% DMEM in 5% CO_2 at 37 °C.
3. BHK$_{NS1}$ cell culture: Grow in 10% DMEM with 0.8 μg/mL of puromycin (diluted in 20 mM HEPES) in 5% CO_2 at 37 °C.
4. Phosphate buffer solution (PBS; 1×): 135 mM NaCl, 2.7 mM KCl, 1.5 mM KH_2PO_4, and 8 mM K_2HPO_4, pH 7.2. Store at room temperature.
5. 0.25% trypsin-EDTA. Store at −20 °C.
6. 24-well plates.
7. Pipettes.

8. 50 mL tubes.

9. Microscope.

10. Carbon dioxide constant temperature incubator (37 °C, 5% CO_2).

11. TC20™ automated cell counter, cell-counting slides (Bio-Rad).

2.3 Plaque Assay

1. 2% DMEM: DMEM medium containing 2% heat-inactivated FBS, 1% penicillin-streptomycin solution. Store at 4 °C.

2. 1% Methylcellulose solution: DMEM medium containing 2% FBS and 1% methylcellulose. Store at 4 °C.

3. 96-well plates.

2.4 Immunostaining

1. Fixation solution: Acetone-methanol (1:1 volume/volume). Store at −20 °C.

2. Staining solution: The staining solution is prepared by mixing 1 mL of 1 × HRP reaction buffer with 50 μL of reagent A, 50 μL of reagent B, and 50 μL of reagent C successively (*see* **Note 1**). The reaction buffer and all reagents are from the enhanced HRP-3, 3′-diaminobenzidine (DAB) chromogenic substrate kit.

3 Methods

3.1 Cell Culture

One day before the experiment, seed 1×10^5 BHK-21 cells or BHK_{NS1} cells per well in a 24-well plate.

1. Discard the culture supernatant of BHK-21 cells or BHK_{NS1} cells in the T-25 culture flask, and wash the cells with PBS once.

2. Discard PBS and add 400 μL of pre-warmed 0.25% trypsin-EDTA (*see* **Note 2**). Swirl the culture flask to cover all the cells with 0.25% trypsin-EDTA.

3. Digest the cells in the incubator for 1 minute at 37 °C (*see* **Note 3**).

4. Tap the culture flask and observe the cells under a microscope to check whether the cells are detached from the culture flask.

5. Add 4–5 mL of pre-warmed 10% DMEM into the flask immediately (*see* **Note 4**), and pipette cells up and down to ensure all cells are in suspension.

6. Transfer the cell suspension into a 50 mL of tube with a pipette and shake gently to mix cells well.

7. Draw 10 μL of cell suspension into cell-counting slide and insert into the TC20™ automated cell counter to quantify the cell number (*see* **Note 5**).

8. Dilute cell suspension with 10% DMEM to a final concentration of 1×10^5 cells/mL and add 1 mL of diluted cell suspension per well into the 24-well plate.

9. Tilt the plate front and back, left and right, to distribute the cells evenly across the plate and incubate for 1 day at 37 °C in a 5% CO_2 incubator.

The following steps should be performed in a BSL-3 laboratory.

3.2 Single-Layer Cell Infection

Before this experiment, check if the cells in the 24-well plate have reached 80%–90% confluence (*see* **Note 6**).

1. Take a vial of virus from −80 °C freezer and thaw it on ice in advance.

2. A series of 1:10 dilutions are made by mixing 15 μL of virus stocks with 135 μL of 2% DMEM in a 96-well plate.

3. Discard the culture supernatant of the 24-well plate (*see* **Note 7**). 100 μL of each dilution is added to individual wells of the 24-well plate (*see* **Notes 8** and **9**).

4. Incubate the plate at 37 °C in 5% CO_2 with manual shaking at 15 min intervals (*see* **Note 10**).

5. One hour later, an overlay of 1 mL of 1% methylcellulose solution is added after the removal of the virus inoculum (*see* **Note 11**), and the plate is incubated further at 37 °C for 24–36 h (*see* **Note 12**).

3.3 Immunostaining

1. Remove the methylcellulose solution, and wash the cells with PBS 3 times (*see* **Note 13**).

2. Add 1 mL of ice-cold fixation solution to each well and incubate for at least 20 min at room temperature (*see* **Note 14**).

3. Discard the fixation solution, and wash the cells with PBS three times.

The remaining steps are not necessarily performed in a BSL-3 laboratory.

4. Add 250 μL of the diluted 4G2 antibody (1.9 mg/mL, 1:500 dilution, diluted in PBS) to each well and incubate for 1 h at room temperature.

5. Discard the primary antibody solution and wash the cells with PBS three times.

6. Add 250 μL of the diluted HRP-conjugated goat anti-mouse IgG secondary antibody (0.2 mg/mL, 1:1000 dilution, diluted in PBS) to each well and incubate for 2 h at room temperature (*see* **Note 15**).

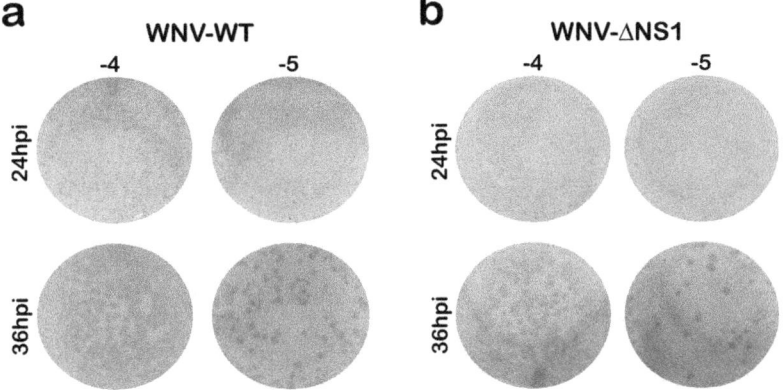

Fig. 1 Plaque morphology of WNV-WT and WNV-ΔNS1 at 24 hpi and 36 hpi. (**a**) Plaque morphology of WNV-WT in 10^{-4} and 10^{-5} dilution obtained by incubating the plate at 37 °C for 24 hpi and 36 hpi, respectively. (**b**) Plaque morphology of WNV-ΔNS1 in 10^{-4} and 10^{-5} dilution obtained by incubating the plate at 37 °C for 24 hpi and 36 hpi, respectively

7. Discard the secondary antibody solution and wash the cells with PBS 3 times.

8. Add 250 μL of staining solution to each well and incubate for 30 min in the dark at room temperature when blue-purple clusters were clear (*see* **Note 16**).

9. Rinse the plate with running tap water to remove the residual staining solutions followed by air-dried, and the plaque morphology is photographed or scanned (Fig. 1). The numbers are counted and viral titer is calculated as below:

Viral titer (foci forming units per milliliter, FFU/mL) = the average number of plaques/(dilution factor × volumes of virus added to the well) (e.g., *see* **Note 17**).

4   Notes

1. The staining solution should be stored away from light and used within 30 min.

2. According to our experience, 400 μL of 0.25% trypsin-EDTA is enough to detach the cells that grow in a T-25 culture flask.

3. The digestion time could be longer if the cells have not been detached completely.

4. 4 ~ 5 mL of 10% DMEM are sufficient to terminate the trypsin digestion.

5. If the TC20™ automated cell counter is not available, draw 10 μL of cell suspension into the hemacytometer to count the cells. The calculation formula is as follows: Number of cells/

mL = total number of cells in the four squares (top left, bottom left, top right, bottom right) of the counting board $\times 10^4$ / 4.

6. We find that 80 ~ 90% confluent cells are optimal for the observation of plaque morphology.

7. Prepare an appropriate waste container, and pour the culture supernatant into the waste container directly. Be careful not to let the culture plate be too dry.

8. If adding the solutions from the highest dilution to the lowest dilution, the same pipette tip can be used repeatedly.

9. The adding process should be handled as fast as possible to avoid drying the wells and poking the cell monolayer with pipette tip.

10. Gently swirl the plates to spread the virus inoculum evenly in the wells.

11. The methylcellulose adding process should be handled as fast as possible to avoid drying the wells and poking the cell monolayer with pipette tip.

12. We find that the plaque morphology is more easily observed at 36 h postinfection (hpi) than 24 hpi (Fig. 1).

13. In order to enable the antibody to bind to the viral antigen effectively, try to remove all traces of the methylcellulose as possible as you can.

14. For better fixation effect, the fixation solution should be stored at $-20\ ^{\circ}C$, and do not take it out until the last wash.

15. We find that diluting the secondary antibody to 1:1000 (v/v) could decrease the background to some extent compared to 1: 500 dilution.

16. The dyeing period varies depending on the color changes.

17. For example, if six plaques are the average number of plaques in the dilution of 10^{-6}, viral titer = six plaques observed/ (10^{-6} dilution factor $\times$ 0.1 mL of viruses added) = 6 $\times$ 10^7 FFU/mL.

Acknowledgments

This work was supported by the National Key Research and Development Program of China (2018YFA0507201) to BZ.

References

1. Kimani J, Osanjo G, Sang R et al (2016) Performance of methylcellulose and Avicel overlays in plaque and focus assays of Chikungunya virus. Afr J Pharm Pharmacol 5(2):54–58

2. Zou G, Zhang B, Lim PY et al (2009) Exclusion of West Nile virus superinfection through RNA replication. J Virol 83(22):11765–11776

3. Li XF, Dong HL, Wang HJ et al (2018) Development of a chimeric Zika vaccine using a licensed live-attenuated flavivirus vaccine as backbone. Nat Commun 9(1):673

4. Hollý J, Fogelová M, Jakubcová L et al (2017) Comparison of infectious influenza a virus quantification methods employing immunostaining. J Virol Methods 247:107–113

5. Harder TC, Klusmeye K, Frey H-R et al (1993) Intertypic differentiation and detection of intratypic variants among canine and phocid morbillivirus isolates by kinetic neutralization using a novel immunoplaque assay. J Virol Methods 41(1):77–92

6. Crameri G, Wang L-F, Morrissy C et al (2002) A rapid immune plaque assay for the detection of Hendra and Nipah viruses and anti-virus antibodies. J Virol Methods 99(1–2):41–51

7. Jorquera PA, Tripp RA (2016) Quantification of RSV infectious particles by plaque assay and immunostaining assay. Methods Mol Biol 1442:33–40

8. Park JS, Um J, Choi YK et al (2016) Immunostained plaque assay for detection and titration of rabies virus infectivity. J Virol Methods 228:21–25

9. Zhang HL, Ye HQ, Deng CL et al (2017) Generation and characterization of West Nile pseudo-infectious reporter virus for antiviral screening. Antivir Res 141:38–47

10. Li N, Zhang Y-N, Deng C-L et al (2019) Replication-deficient West Nile virus with NS1 deletion as a new vaccine platform. J Virol 93(17):e00720–e00719

11. Shi PY, Tilgner M, Lo MK et al (2002) Infectious cDNA clone of the epidemic west nile virus from New York City. J Virol 76(12): 5847–5856

12. Zhang PT, Shan C, Li XD et al (2016) Generation of a recombinant West Nile virus stably expressing the Gaussia luciferase for neutralization assay. Virus Res 211:17–24

Chapter 4

Isolate and Culture Mouse Primary Neurons for West Nile Virus Infection

Farzana Nazneen and Fengwei Bai

Abstract

Primary neurons are very valuable cells to study the pathogenesis of neurotropic viruses, such as West Nile virus. The mouse primary neurons can be used to assess viral infection profiles and cellular immune responses to a viral infection. However, successful isolation and culture of mouse neurons can be challenging. Here, we report a step-by-step method to prepare a primary neuron culture from adult mice.

Key words Mouse primary neuron, Culture protocol, Neurotrophic virus, West Nile virus

1 Introduction

Primary neurons are directly isolated from animal brain to study the physiological and pathological impacts of any insults caused by bacteria, virus, fungus, or toxins. Both embryonic, postnatal and adult neuronal cells can be used to establish an in vitro primary neuron culture model [1]. Mouse embryonic neuron cells can be collected from embryonic days (E) 11 to E17 [2]. Embryonic neuron cells are robust to sustain the stress during the cell isolation and collection procedure. Although neuron cells can also be collected at any postnatal age, mice between 4- and 7-week old are preferred. In this chapter, we present a technique of collection of mouse brain from 7-week-old mice and culture the neuron cells to establish the primary neuron culture model for any downstream studies.

As neuron and non-neuron cells such as glia act differently upon a viral infection, a protocol that isolates pure neurons is warranted for successful downstream analysis. There are some commercially synthesized components available that can separate the neuron cells from non-neuronal cells while culturing the neuron cells, such as poly-l-ornithine (PLO), poly-D-lysin (PLL), and fibronectin (FN). They act as the adhesive molecule for neuron.

Fengwei Bai (ed.), *West Nile Virus: Methods and Protocols*,
Methods in Molecular Biology, vol. 2585, https://doi.org/10.1007/978-1-0716-2760-0_4,

When the culture plates are coated with these coating materials, only the neuron cells can attach to the surface, and the rest of non-neuron cells are washed away. Moreover, specific types of plating and feeding medium are used to support growth of only neuron cells. The coating reagents also promote proliferation and differentiation of neuron stem/progenitor cells (NSPCs). Studies show that PLO significantly increases the proliferation and differentiation of NSPCs compared to PLL and FN [3]. Moreover, PLO is less immunogenic to cells than PLL [4]. We have used PLO-coated plates to culture the primary neuron cells collected from adult mice. Typically, an adult mouse brain contains 70 million neurons [5]. Careful and gentle handling during the cell isolation process can ensure a yield of 50% of neuron cells.

2 Materials

2.1 Collection of Brain from Mice

1. Surgical set containing forceps, fine scissors, coarse scissors, and tweezers.
2. 30% isoflurane in Isopropyl alcohol.
3. Phosphate buffer saline (PBS).
4. 50 mL syringe.
5. Needles (20G).
6. Butterfly needle (21G).
7. Dissection station or Table.
8. 70% Ethanol.
9. 10% Bleach.
10. Styrofoam.
11. Tray.

2.2 Processing the Cells

1. HEPES buffer saline (HBS): Dissolve 4.24 g of NaCl (145 mM), 820.16 mg of KCl (22 mM), and 450.5 mg of D-(+)-glucose (5 mM) in 500 mL nano-pure water (NP H_2O), and autoclave the solution. Add 10 mM of HEPES solution (5 mL); adjust the pH to 7.3. Store at 4 °C (see **Note 1**).
2. 2 mg/mL papain solution: Mix papain powder with HBS solution for the final concentration of 2 mg/mL, and add 50 units/mL of DNase I, RNase-free (1 unit/μL, Thermo Scientific™). Store at 4 °C (see **Note 2**).
3. 50 μg/mL poly-l-ornithine (PLO) solution: Dilute 0.1 mg/mL PLO solution (Sigma-Aldrich) 1:1 (v/v) in NP H_2O or PBS. Store at −20 °C.
4. 70 μm cell strainers.
5. Six-well plates.

6. Plating medium: Add Dulbecco's Modified Eagle Medium/Nutrient Mixture F-12 (1:1) (133.5 mL), 0.5625 g of D-(+)-glucose, 10% fetal bovine serum (15 mL), and 1% penicillin-streptomycin (5000 U/mL, 1.5 mL). Filter the solution with vacuum filtration systems with 0.2 μm pore size. Store at 4 °C (*see* **Note 3**).

7. Feeding medium: Add Neurobasal™ Plus Medium (Gibco™, 500 mL), B-27™ Plus Supplement (50×, Gibco™, 10 mL), and 1% penicillin-streptomycin (5000 U/mL, 5.1 mL). Filter the solution with vacuum filtration systems with 0.2 μm pore size. Store at 4 °C.

3 Method

3.1 Day Before the Collection of Mice Brain (D-1)

1. Coat culture surface of six-well plates with 2 mL/well PLO solution. Incubate the plates at room temperature (RT) overnight (*see* **Note 4**).

2. Prepare 1 mL of papain solution per mouse. Incubate at 37 °C overnight.

3. Autoclave the surgical tools needed for the dissection of mice.

3.2 Day of Collection and Processing the Cells (D0)

1. Discard PLO solution from the plates, and wash three times with sterile 4 mL/well NP H_2O in six-well plates.

2. Remove excess NP H_2O from the plates, and leave the plate uncovered in a running hood to dry for 2 hours (h) (*see* **Note 5**).

3.3 Collection of Neuron Cells from Mice (D0)

1. Aliquot 5 mL of HBS solution and 20 mL of PBS per mouse, and keep chilled on ice. Put PBS in 50 mL syringe attached with a butterfly needle.

2. Clean the workstation and necessary tools for dissection with 70% ethanol, and prepare a beaker with 70% ethanol to keep the surgical sets in between use (*see* **Note 6**).

3. Attach a Styrofoam tightly in a tray where the animal will be dissected, and add 10% bleach in the tray to collect blood of the animal and other liquid waste products.

4. Perfusion of mice: Put a mouse into a jar which contains a few drops of 30% isoflurane solution in the bottom and covered with a piece of paper tower and close the jar lid (*see* **Note 7**). Monitor the respiration rate of the animal. Once the respiration rate drops to 2–3 breaths/min, start the perfusion procedure immediately. Fasten the animal face up with pining down the four limbs onto a piece of Styrofoam board over the liquid collecting tray. Cut the skin of the thorax with course scissors, expose the chest region, and then cut several ribs of the left side of the chest to expose the heart. Insert the butterfly needle into

the apex of the heart (tip of the left ventricle), and inject 20 mL of PBS per mouse through the apex gently (*see* **Note 8**).

5. Remove the head from the neck, cut the skin on the top of the skull using fine scissors, and peel off the skin from the head (*see* **Note 9**).

6. Using the fine scissors tip, puncture the skull in middle of the lower margin of the occipital bone, point A. Start cutting the skull in a semicircular manner using fine scissors from point A along the lower margin of the occipital bone until the beginning of ear, point C on both sides. Make a vertical cut starting from point A along the sagittal suture line till the end of the nasal bone, point B. The sagittal suture line is a clearly visible grove passing between the parietal, frontal, and nasal bones (Fig. 1) (*see* **Note 10**).

7. Remove the bones from the skull using tweezers. Now the brain can be easily taken out from the base of the skull using

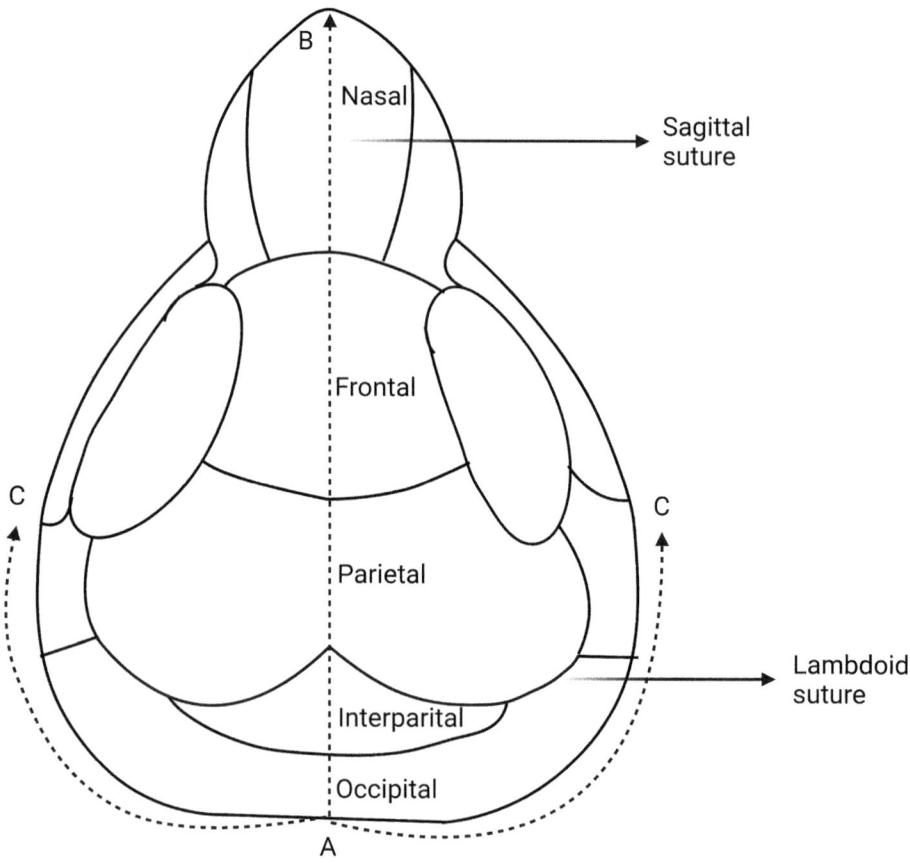

Fig. 1 The skull of the mouse brain showing the incision lines to open up the brain. Puncture the skull at the midpoint of the occipital bone, point A; make a semicircular cut along the lower margin of the occipital bone up to point C on both sides of the skull; and make another vertical cut starting from point A to point B. The illustration was created with BioRender

forceps. Cut the spinal cord from the brain with scissors. Put the brain in a 15 mL centrifuge tube containing 5 mL ice-cold HBS solution (*see* **Note 11**).

3.4 Processing the Neuron Cells (D0)

1. Take out the brain from the centrifuge tube, put it in a sterile 10 mm petri dish, and mince the brain with a sterile rubber plunger of a 5 mL syringe or autoclaved blades. Put the minced brain tissue back to the centrifuge tube containing HBS, and let it settle down for 20 minutes (min) at RT.

2. Gently remove HBS without disturbing the cells. Resuspend the minced brain tissue in 1 mL of papain solution per/brain. Incubate for 15 min at 37 °C, and mix every 5 min by inverting the tube gently.

3. Add 4 mL of plating medium per brain into the minced brain tissue. Triturate the mix 30 times with 1 mL pipette until nearly cloudy (*see* **Note 12**).

4. Filter the cell medium mix through a 70 μm cell strainer. Use a plunger of 5 mL syringe to push the remaining chunks of brain tissue. Add 1 mL of plating medium per brain into the strainer to facilitate the filtration.

5. Centrifuge the filtrate at $530\times$ g for 5 min at 4 °C. Discard the supernatant.

6. Add 5 mL of pre-warmed plating medium per brain and mix gently.

7. Count the cells with hemocytometer using a microscope. Dilute the cells five- to tenfold (x) with plating medium and then $2\times$ with 0.4% trypan blue. The viability of the obtained cells can be >50%.

8. Dilute the neuron cells at 1×10^6 cells/mL with plating medium pre-warmed at 37 °C (*see* **Note 13**).

9. Plate 2×10^6 neuron cells in PLO-coated six-well plates (pre-prepared).

10. Incubate the neuron cells at RT for 25 min and then at 37 °C overnight in a humidified incubator with 5% CO_2.

3.5 Processing the Neuron Cells (D1)

1. Observe the cells under an inverted microscope; if attached, remove medium, wash the cells with 2 mL/well pre-warm HBS, and add 2 mL/well pre-warm Plating medium.

2. Incubate the cells at 37 °C for 4 days.

3.6 Maintaining the Neuron Cells

1. Observe the growth of the cells every day. On D4, replace 1 mL medium from the wells with 1 mL/well pre-warmed feeding

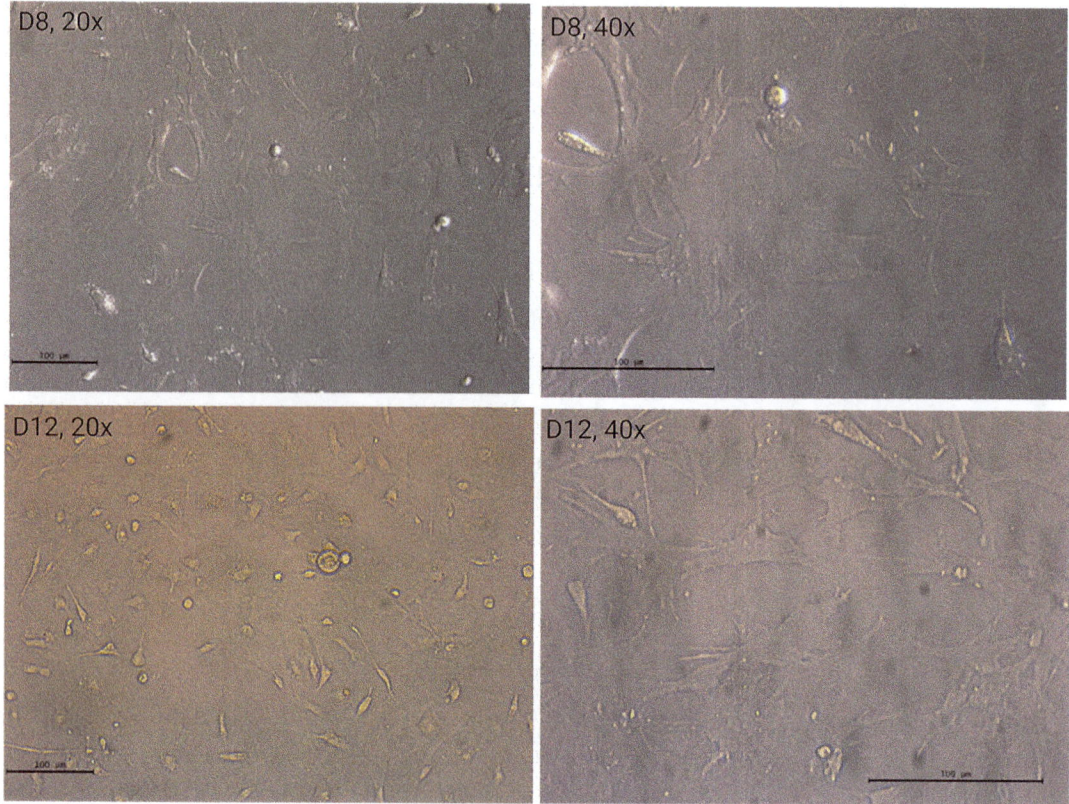

Fig. 2 Images of primary neuron cells taken on 20×, and 40× magnification of D8 and D12 of plating using Leica M165 FC microscope. The illustration was created with BioRender

medium. In every 4 days, replace 1 mL medium with 1 mL/well pre-warm feeding medium (*see* **Note 14**).

2. The neuron cells can be grown up to 21 days with regular changing of feeding medium. By D11, the cells should be ready for experiment (Fig. 2).

4   Notes

1. While preparing the solution, add NaCl, KCl, and glucose powder in NP H_2O; mix; and autoclave. Then add the HEPES solution in the mixture. If needed, HBS solution can be filtered before use. Instead of HEPES solution, HEPES powder can also be used. In that case, autoclave HEPES powder before adding to the solution.

2. Papain is supplied as either suspension or lyophilized powder. Dilute papain suspension or powder with HBS solution to the

recommended concentration. The enzyme suspension should be incubated at 37 °C for at least 30 min to ensure full enzymatic activity. In our protocol, we recommended incubation of papain solution overnight. It is also necessary to filter the enzyme suspension with 0.22-micron filter prior to use. Papain solution is stable at 2–8 °C for 6–12 months. But don't freeze the aqueous suspension.

3. First, autoclave the required amount of glucose powder, then add the remaining ingredients, mix, and filter.

4. PLO solution helps the neuron cells to attach to the culture surface. PLO solution can be diluted by either NP H_2O or PBS. Here, we diluted the PLO solution with NP H_2O to a final concentration of 50 µg/mL. Typical coating concentrations can range from 10 µg/mL to 100 µg/mL. Use 5 mL of PLO solution in 6 cm plates and 10 mL for 10 cm plates and T75 flasks. Store the PLO solution at −20 °C up to 4 months. Before starting the procedure, thaw the PLO solution at RT. Avoid multiple freeze thaw cycles to maintain integrity of the product.

5. Rinse the plates thoroughly with NP H_2O as PLO solution can be toxic to the cells [6]. PBS can also be used to wash the plates. Once dry, plates can be used immediately or stored for 1 week at 4 °C. For storage, tightly wrap the plates with Parafilm.

6. Always put the seizers, tweezers, and forceps submersed in 70% ethanol beaker prepared specifically for this step to maintain sterility.

7. Before starting any animal procedure, an approved IACUC protocol of animal handling is required.

8. Try to keep the mice alive until this point. If the heart stops beating, the PBS will not enter into the brain. Inject PBS solution into the apex of the heart slowly. It will give enough time to circulate PBS inside the body as well as the brain. If the PBS is properly infused into the brain, the brain will turn into chalky white in color. The success of primary neuron culture depends on how efficiently the brain is infused with PBS. You can also try another technique for PBS infusion. Make a small hole in the upper margin of the right side of the heart with fine seizers, and then inject PBS into the apex of the heart. In this case, PBS will not circulate back in the left heart after infusing into the brain. It will ensure better perfusion of PBS inside the brain.

9. Completely remove the skin from the skull from all over the head. Wash the skull with PBS to make sure no hair is left in the

skull. Careful removal of hair from skull is necessary as it can contaminate the culture of neuron tissue. If the incisions/cuts are deep enough, you can easily pull out the bones from the skull.

10. Try to cut through the skull as deep as possible from point A to point C. The end point of point C can be up to the beginning of the ear. But in the case of cutting through the point A to point B, you need to cut superficially. A deep cut can separate the cerebral cortexes of the mice brain. You can cut the tip of the nasal portion which helps in opening up the skull easily.

11. The mice brain should always be kept in ice until the end of plating. To ensure high yield in live neuron cells, the processing time should be as less as possible.

12. Triturate the cell mix very gently. Because rough handling can damage the neuronal cells. Fire-polished Pasteur pipettes can also be used to triturate the cell mix.

13. Plating media must be pre-warmed at 37 °C in water bath. It ensures good yield in neuron culture.

14. The medium in which the primary neuron cells are growing contains necessary growth factors, so don't replace the whole media from the plates. Always change half of the medium from the plates. Always use pre-warmed feeding media to replenish the plates.

Success in culturing primary neuron mainly depends on the handling of the brain cells. Gentle handling can assure good yield in cell number and viability. As the neuron cells need to be grown for longer period of time in the same culture plates/wells, it is recommended to maintain strict aseptic technique to avoid contamination.

Acknowledgments

This work was supported in part by the National Institute of Allergy and Infectious Diseases of the National Institutes of Health R15AI135893 (F.B.).

References

1. Facci L, Skaper SD Culture of rodent cortical and hippocampal neurons. (1940–6029 (Electronic))

2. Sciarretta C, Minichiello L The preparation of primary cortical neuron cultures and a practical application using immunofluorescent cytochemistry. (1940–6029 (Electronic))

3. Ge H, Tan L, Wu P et al (2015) Poly-L-ornithine promotes preferred differentiation of neural stem/progenitor cells via ERK signalling pathway. Sci Rep 5:15535–15535. https://doi.org/10.1038/srep15535

4. Brunetti P, Basta G Faloerni A, Calcinaro F et al Immunoprotection of pancreatic islet grafts within artificial microcapsules. (0391–3988 (Print))

5. Herculano-Houzel S, Mota B, Lent R (2006) Cellular scaling rules for rodent brains. Proc Natl Acad Sci 103(32):12138. https://doi.org/10.1073/pnas.0604911103

6. Calvo-Garrido J, Winn D, Maffezzini C et al (2021) Protocol for the derivation, culturing, and differentiation of human iPS-cell-derived neuroepithelial stem cells to study neural differentiation in vitro. STAR Protoc 2(2):100528–100528. https://doi.org/10.1016/j.xpro.2021.100528

Chapter 5

Isolation of Murine Bone Marrow-Derived Neutrophils for Infection Modeling

Laurel Duty and Amber M. Paul

Abstract

West Nile virus (WNV) is a single-stranded-RNA flavivirus that can cause neurological illnesses. The ability of WNV to cause neurological illnesses is dependent on the virus' ability to gain access into the brain. However, the mechanisms by which WNV enters the brain are elusive, with evidence that neutrophils act as vehicles for viral delivery. To determine the role of neutrophils in WNV central nervous system delivery, modified protocols for the isolation and migration of neutrophils from mouse bone marrow for in vitro WNV infection modeling is described.

Key words Bone marrow-derived neutrophils, West Nile virus, Migration, Osteopontin

1 Introduction

West Nile virus (WNV) is a single-stranded (ss)-RNA flavivirus belonging to the family *Flaviviridae*. Its transmission cycle involves amplification of the virus within an avian reservoir and is primarily transmitted to humans by the bite of an infected mosquito. Most human cases are asymptomatic (80%), while others show febrile illness-like symptoms. Less than 1% of individuals develop severe neurological illnesses, including meningitis, encephalitis, and flaccid paralysis [1]. Therefore, WNV has been considered as a neurotropic virus, as it can readily infect neurons within the central nervous system (CNS). Although the mechanisms of viral entry into the CNS are not completely understood, three mechanisms have been proposed: [1] direct entry through the blood-brain barrier (BBB), [2] retrograde transport of the peripheral nervous system (PNS), and [3] a "Trojan horse" mechanism, whereby the virus hijacks immune cells to enter into the CNS [2]. There are currently no specific antiviral vaccine or therapeutic treatment available for human WNV infection [3].

Fengwei Bai (ed.), *West Nile Virus: Methods and Protocols*,
Methods in Molecular Biology, vol. 2585, https://doi.org/10.1007/978-1-0716-2760-0_5,

Since neutrophils are the most abundant leukocytes in blood and are described as important viral reservoirs for WNV [4], we propose neutrophils may be a chief cell type that deliver WNV into the CNS by the "Trojan horse" mechanism [2]. An important biological function of neutrophils is to phagocytize foreign material [5] and is recruited by various chemokines. Therefore, to determine the role of neutrophils in WNV-CNS delivery, a modified protocol for the isolation and migration of murine bone marrow-derived neutrophils is described.

2   Materials

For successful implementation, an efficient protocol that can produce viable neutrophil populations, without contamination of other leukocytes, is necessary. In addition, a protocol that isolates neutrophils from the bone marrow rather than isolation from blood is preferred, as bone marrow-derived neutrophils are less mature than blood derived neutrophils, making them suitable for subsequent activation and infection modeling [6]. Swamydas et al. previously reported isolation protocols for murine bone marrow-derived neutrophils based on density gradient centrifugation [7, 8]. Herein, we describe these neutrophil isolation protocols with some modifications along with an additional modified migration protocol [9] for neutrophil "Trojan horse" modeling. These protocols were used in our recent report that identified neutrophils as vehicles for viral delivery into the CNS [10].

Unless indicated otherwise, all solutions require ice-cold temperatures to maintain viability of isolated neutrophils and majority of presented materials used in this protocol were purchase from Thermo Fisher Scientific. Listed volumes are designed to accommodate two femur bone marrow isolations from one mouse. All animal experimental procedures were reviewed and approved by the Institutional Animal Care and Use Committee (IACUC) at the University of Southern Mississippi (protocol #12041201).

2.1   Isolation of Murine Bone Marrow-Derived Neutrophils

1. 20 mL of 70% ethanol in water.

2. Plating media: RPMI 1640, 10% fetal bovine serum (FBS) and 1% penicillin/streptomycin (Pen/Strep).

3. Two, 10 mm deep tissue culture Petri dishes.

4. Dissection forceps.

5. Dissection scissors.

6. Kimwipes™.

7. 10 mL syringe.

8. 25 gauge needle.

9. 10 mL serological pipettes.

10. 1 mL serological pipettes.

11. A handheld pipettor.

12. 50 mL and 15 mL conical tubes.

13. 1.5 mL tubes.

14. $1 \times$ phosphate buffered saline ($1 \times$ PBS, without calcium and magnesium).

15. 0.2% NaCl: Add 0.2 g of NaCl in 100 mL of filter sterilized water. Use 10 mL of aliquots for experiment, and store remaining in 4 °C for up to 1 month.

16. 1.6% NaCl: Add 1.6 g of NaCl in 100 mL of filter sterilized water. Use 10 mL of aliquots for experiment and store remaining in 4 °C for up to 1 month.

17. Histopaque®-1077 (density: 1.077 g/mL, Sigma-Aldrich) and Histopaque®-1119 (density: 1.119 g/mL, Sigma-Aldrich).

18. Refrigerated centrifuge with no brake option.

19. 70 μm cell strainer.

20. Flow cytometer.

21. Antibodies to determine neutrophil purity. For this, anti-CD11b, anti-CD45, and anti-Ly6G antibodies conjugated to appropriate fluorophores are required for neutrophil identification (*see* **Note 1**).

2.2 Neutrophil Migration Assay

1. Recombinant mouse osteopontin (rOPN, cat. no. 441-OP-050, R&D Systems) (*see* **Note 2**).

2. 12 mm, 12-well Transwell® plates with 3 μm pore polyester membrane inserts (cat. no. 3462, Corning), 12-well cell culture plates, and 1.5 mL tubes.

3. 4′,6-Diamidino-2-phenylindole (DAPI) DNA fluorescent stain.

4. 4% Paraformaldehyde (PFA) in $1 \times$ PBS.

5. RPMI 1640.

6. $1 \times$ PBS (without calcium and magnesium).

7. Standard unrefrigerated centrifuge.

8. Confocal or fluorescent microscope and a microplate fluorometer with DAPI detection capabilities (UV light and blue/cyan filter).

3 Methods

3.1 Isolation of Murine Bone Marrow-Derived Neutrophils

1. A 7- to 9-week-old mouse is CO_2 euthanized following IACUC guidelines. Mouse is saturated in 70% ethanol and both femurs are removed. Muscles are immediately separated from femurs using Kimwipes™.

2. Femurs are disinfected in ice-cold 70% ethanol (up to 30 s) and are placed in a 10 mm deep tissue culture dish containing 10 mL of ice-cold 1× PBS.

3. Isolated femurs are flushed with 10 mL of 4 °C plating media in a 10-mm-deep tissue culture plate.

4. Use dissection scissors and forceps to carefully remove femur epiphyses and a 10 mL syringe and 25-gauge needle to flush out hematopoietic bone marrow cells (*see* **Note 3**).

5. Hematopoietic bone marrow cells are dissociated gently with 1-mL serological pipetting (approximately 1 min in duration) to homogenate cells followed by centrifugation at 430 × g for 7 min in a 4 °C refrigerated centrifuge (*see* **Note 4**).

6. Decant supernatant and the remaining pellet should be visible. Lyse red blood cells (RBC) using 10 mL of 0.2% NaCl (20 s) and 10 mL of 1.6% NaCl (20 s) to stepwise lyse RBC.

7. Add 10 mL of 4 °C plating media to lysed cell solution to ensure lysing reaction is isotonic buffered.

8. Pass cells through a 70 μm cell strainer to remove debris from lysed cells.

9. Centrifuge cells at 430 × g for 7 min in 4 °C centrifuge.

10. Wash cells with 10 mL of 4 °C plating media once, and centrifuge at 430 × g for 7 min in 4 °C.

11. Cells are resuspended in 1 mL of ice-cold 1× PBS and stored on ice until layered on top of Histopaque® solutions (**steps 12** and **13**).

12. In a 15 mL conical tube, slowly added 3 mL of room temperature Histopaque®-1119 followed by slowly adding 3 mL of room temperature Histopaque®-1077 on top (*see* **Notes 5** and **6**).

13. Slowly add 1 mL of ice-cold 1× PBS containing cells to the top of the Histopaque®-1077 layer.

14. Immediately centrifuge at 870 × g for 30 min without the break in a room temperature centrifuge.

15. Post-centrifuge neutrophils are located within the interface of the Histopaque®-1077 and Histopaque®-1119 layers and is visually observed as a cloudy ring (Fig. 1a, *see* **Note 7**).

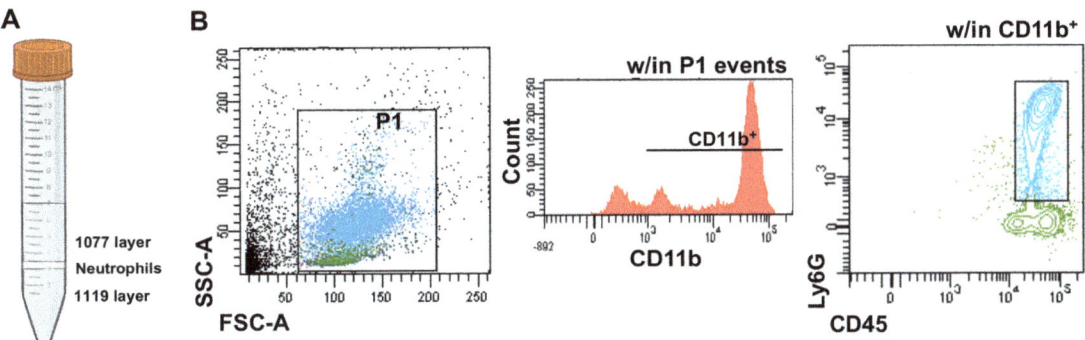

Fig. 1 Isolation of bone marrow-derived neutrophils. (**a**) Image depict interface layer of neutrophils post-density gradient centrifugation with Histopaque®-1119/−1077 layering. Cloudy interface between 1077 and 1119 layers (near 3 mL mark on 15 mL conical tube) is locale for neutrophils. (**b**) Scatter plot of population one (P1) events collected within isolated neutrophil layer. Blue events denote neutrophils (CD11b⁺ w/in CD45⁺Ly6G⁺ events). Yielded collections depict >75% neutrophil purity. Figure 1a was created in BioRender

16. Carefully collect this interface layer of neutrophils, and wash with 10 mL of room temperature plating media twice, centrifuging at $430 \times g$ for 7 min in a room temperature centrifuge.

17. Resuspend neutrophils in 37 °C pre-warmed plating medium. This yields approximately $1–5 \times 10^6$ cells/mL with a purity of >75% (Fig. 1b, *see* **Note 8**) neutrophils and can be used immediately for subsequent assays (*see* **Note 9**).

3.2 Neutrophil Migration Assay

To assess neutrophil migration, a transwell migration assay is applied using isolated, uninfected neutrophils, as described above. This migration protocol is adapted from Koh et al. with some modifications [9]. Previous work in our lab has identified OPN to be an important chemokine involved in neutrophil recruitment and is involved in WNV neuroviral pathogenesis [10]. Herein, we describe a modified neutrophil migration protocol that utilizes recombinant mouse OPN (rOPN) as the chemotactic agent to model neutrophil migration. Unless otherwise indicated, all centrifugation cycles are performed at $430 \times g$ for 7 min in room temperature, and neutrophils are kept at optimal cell growth conditions of 37 °C and 5% CO_2.

1. Twelve-well, 12 mm transwell plates with 3 μm pore membrane inserts are used for the neutrophil migration assay (Fig. 2a).

2. 600 μL of RPMI 1640 is added to the bottom chamber of each well followed by the addition of murine rOPN at 2 μg/mL or media only controls at similar volumes.

3. 3 μm pore membrane inserts are added to each well.

4. 8×10^5 bone marrow-derived neutrophils are isolated as described above and resuspended in 400 μL of RPMI 1640.

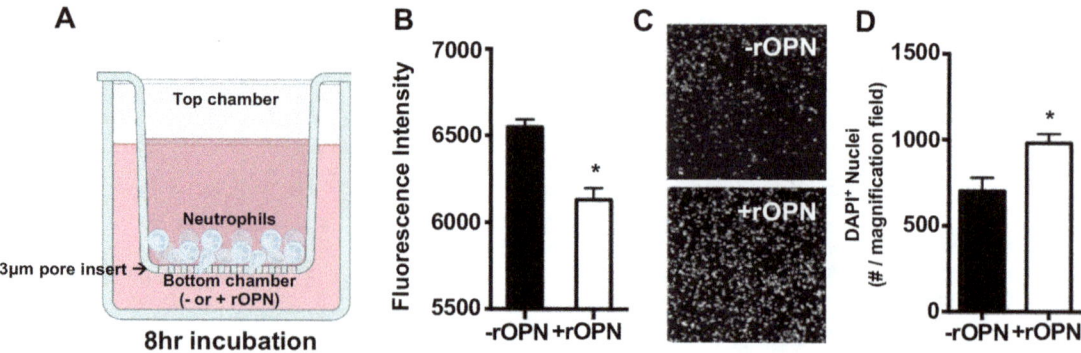

Fig. 2 Neutrophil migration assay. (**a**) Representative image of transwell with 3 μm pore inserts. (**b**) Pre-stained with DAPI the fluorescence intensity corresponds to the number of neutrophils remaining in the top transwell. (**c**) The transwell containing interface membrane with (+) and without (−) recombinant osteopontin (rOPN). (**d**) Quantification of DAPI$^+$ nuclei of neutrophils visible in the field of view at 10× magnification ($p < 0.05$). Data were compared with a Student's t-test. All statistical analysis were completed using GraphPad Prism software (version 6.0). Figure 2d is from Paul et al. [10] with changes to text fonts within the image and used with open access, a creative common permission license (https:/creativecommons. org/licenses/by/4.0/). Figure 2a was created in BioRender

5. Cells are directly added to the top chamber of the transwell containing the 3 μm pore inserts.

6. Cells are incubated at 37 °C and 5% CO_2 for 8 h.

7. Following incubation, top chamber containing cells (approximately 350–400 μL) are collected into separate 12-well tissue culture plates and fixed in 2% PFA (from 4% PFA stock) for 15 min at 4 °C.

8. Fixed cells are collected into 1.5-mL tubes and centrifuged for 430 × g for 7 min in room temperature and resuspended in 500 μL of 1× PBS.

9. Cells are stained with DAPI at 1 μg/mL and imaged with a microplate fluorometer to detected DAPI emission (Fig. 2b, *see* **Note 10**).

10. Transwell pore inserts are removed from the incubation media plates and are added to a new 12-well plates. Cells are gently washed once with 1× PBS and fixed in 4% PFA for 15 min at 4 °C (*see* **Note 11**).

11. After incubation, 4% PFA is removed by careful aspiration, and 1× PBS is added to the insert to prevent drying out (*see* **Note 12**).

12. Transwells are imaged under a microscope (Fig. 2c, d) and neutrophils are counted and recorded per magnification field (average of three individual magnification fields).

4   Notes

1. Antibody fluorophores are selected to cytometer capabilities. For optimal results choose antibody fluorophores with least spectral overlap.

2. We test rOPN to determine its potential to act as a chemokine for neutrophils our this report [10]; however, CXCL1 or CXCL2 may also be used as alternatives for neutrophil migration assays.

3. Complete removal of hematopoietic cells from bone marrow is indicated when bone is clear/translucent.

4. Gentle mechanical cell disruption through pipetting to limit nonspecific neutrophil activation.

5. Layering Histopaque®-1077 on top of Histopaque®-1119 needs to be done with careful attention. It must be slowly to avoid layer disruption that can significantly reduce neutrophil yield. Use a 1 mL serological pipette and allow the solutions to run down the inside of the 15 mL conical tube slowly, providing optimal separation of layering.

6. Histopaque® layers must be prepared immediately post-cell collection and not beforehand. Histopaque® layering merge with time, resulting in suboptimal neutrophil isolation if prepared before cell isolation.

7. The Histopaque®-1119/−1077 interface containing neutrophils is found around the 3 mL mark within a 15 mL conical tube.

8. Pooling of different biological samples can reduce neutrophil yield; therefore, it is recommended to perform each protocol with individual mice to maintain optimal yield.

9. Isolated neutrophils are permissive to WNV infection and migration assays [10].

10. Include unstained control cells to determine background fluorescence levels.

11. Careful attention to forces produced from washing steps to not dislodge cells from membrane. Recommend to touch the tip of the pipette to the side of the insert wall, and slowly dispense 1×PBS, slowly layering over insert.

12. Washing step decant was performed by careful pipette aspiration.

Acknowledgments

This work was supported by Embry-Riddle Aeronautical University's start-up fund and Wessel Foundation endowment to AMP.

References

1. Valiakos G, Athanasiou L, Touloudi A et al (2013) West Nile virus: basic principles, replication mechanism, immune response and important genetic determinants of virulence. In (Ed.), Viral Replication. IntechOpen. https://doi.org/10.5772/55198

2. Donadieu E, Bahuon C, Lowenski S et al (2013) Differential virulence and pathogenesis of West Nile viruses. Viruses 5(11):2856–2880

3. Krishnan MN, Garcia-Blanco MA (2014) Targeting host factors to treat West Nile and dengue viral infections. Viruses 6(2):683–708

4. Bai F, Kong KF, Dai J et al (2010) A paradoxical role for neutrophils in the pathogenesis of West Nile virus. J Infect Dis 202(12): 1804–1812

5. Burg ND, Pillinger MH (2001) The neutrophil: function and regulation in innate and humoral immunity. Clin Immunol 99(1):7–17

6. Boxio R, Bossenmeyer-Pourié C, Steinckwich N et al (2004) Mouse bone marrow contains large numbers of functionally competent neutrophils. J Leukoc Biol 75(4):604–611

7. Swamydas M, Lionakis MS (2013) Isolation, purification and labeling of mouse bone marrow neutrophils for functional studies and adoptive transfer experiments. J Vis Exp 77: e50586

8. Swamydas M, Luo Y, Dorf ME et al (2015) Isolation of mouse neutrophils. Curr Protoc Immunol 110: 3.20.1–3.20.15

9. Koh A, da Silva AP, Bansal AK et al (2007) Role of osteopontin in neutrophil function. Immunology 122(4):466–475

10. Paul AM, Acharya D, Duty L et al (2017) Osteopontin facilitates West Nile virus neuroinvasion via neutrophil "Trojan horse" transport. Sci Rep 7(1):4722

Methods to Study West Nile Virus Infection and the Virus-Induced Inflammation in the Brain in a Murine Model

Huanle Luo and Tian Wang

Abstract

West Nile virus (WNV), a mosquito-borne neurotropic flavivirus, has become the leading cause of vector-borne viral encephalitis in the United States for the past decades. The murine model of WNV infection is an effective in vivo experimental model to investigate WNV neuropathogenesis in humans. Here, we describe several laboratory protocols to study WNV infection and the virus-induced inflammation in the brain in both in vitro and in vivo murine models.

Key words West Nile virus, Neuropathogenesis, Inflammation

1 Introduction

West Nile virus (WNV), a mosquito-borne neurotropic flavivirus, has become the most common cause of vector-borne viral encephalitis in the United States over the past two decades. Although the majority of human infections are asymptomatic, about 20% of acute WNV infection in humans present with a spectrum of signs and symptoms ranging from mild flu-like febrile illness, to severe neurological diseases, such as meningitis, encephalitis, acute flaccid paralysis, and death. Up to 50% of recovered patients were reported to have long-term neurological sequelae [1–3]. The elderly and immunocompromised people are known to have a greater risk to WNV-induced severe neurological diseases [4, 5]. Animal models are effective in vivo experimental models to investigate WNV neuropathogenesis [6, 7]. Following systemic infection in mice, WNV enters the central nervous system (CNS), replicates in the brain, and induces inflammation and neuronal death [8–11]. In vitro cell culture studies also suggest that the main CNS residential cells, such as neurons, and glial cells are susceptible to WNV infection

Fengwei Bai (ed.), *West Nile Virus: Methods and Protocols*,
Methods in Molecular Biology, vol. 2585, https://doi.org/10.1007/978-1-0716-2760-0_6,

[12, 13]. Inflammatory responses induced by WNV-infected residential cells [10, 14, 15] and infiltrating cells both contribute to WNV-induced neurological diseases [16–20]. Here, we describe the standard methods used in our laboratory to evaluate WNV infection and the virus-induced inflammatory immune responses in the CNS.

2 Materials

2.1 Isolation of Primary Microglial Cells and Neurons

1. 6- to 10-week-old C57BL/6 (B6) mice.

2. Microglial culture medium: DMEM-F12 medium containing 10% fetal bovine serum (FBS), 1% penicillin/streptomycin (P/S), and 1% GlutaMax.

3. Brain digestion medium: Hanks' Balanced Salt solution (HBSS) containing Ca^{2+} and Mg^{2+}, 10 mM HEPES, 0.6% glucose, and 1% P/S.

4. Neurobasal medium containing 2% B27 supplement, 1% P/S, and 1% GlutaMax.

5. Neurobasal medium.

6. Phosphate-buffered saline (PBS).

7. 10 μg/mL Poly-D-lysin (PDL) diluted in PBS.

8. 2.5% Trypsin (10×).

9. 0.05% Trypsin (1×).

10. Mouse Neuron Isolation Kit (including anti-Biotin Microbeads).

11. Mouse FcR blocker (anti-mouse CD16/32).

12. Perm/wash buffer.

13. Cytofix/cytoperm.

14. Anti-NeuN antibody.

15. Anti-Rabbit IgG-PE.

16. FACS buffer: PBS with 2% FBS.

17. DNase I.

18. Custom-made trypsin solution: 0.25% trypsin, 1 mM of EDTA in HBSS with Ca^{2+} and Mg^{2+}.

19. 0.5% Paraformaldehyde (PFA).

20. 10 cm cell Culture petri dish.

21. 70 μM and 100 μM Cell strainers.

22. LS column.

23. Magnetic multistand separator.

24. 15 mL Conical tube.

25. 12-Well plate.

26. 1.7 mL Microcentrifuge tube.

2.2 Focus-Forming Assays (FFA)

1. Vero cell culture medium: MEM containing 8% FBS and 1% P/S.

2. PBS-gelatin (PBSG): PBS containing 0.1% gelatin.

3. Methylcellulose overlay: MEM containing 0.8% methylcellulose, 3% FBS, 1% P/S, and 1% L-glutamine.

4. PBS.

5. Methanol.

6. Acetone.

7. Blocking buffer: 3% FBS in PBS.

8. Rabbit anti-WNV polyclonal antibody (World Reference Center for Emerging Viruses and Arboviruses (WRCEVA) at UTMB, (#T35502)).

9. Goat anti-rabbit IgG HRP.

10. AEC developing solution.

11. 24-Well plate.

2.3 WNV Infection in Primary Cells and in Mice

1. WNV NY385-99 ([21]): It was obtained from the WRCEVA at UTMB and was passaged once in Vero cells and twice in C6/36 cells to make a virus stock (3.5×10^7 PFU/mL) for all infection studies.

2.4 Isolation of Brain Leukocytes

1. Percoll.

2. RPMI 1640.

3. FBS.

4. PBS.

5. Antibodies: anti-mouse CD3, anti-mouse CD4, anti-mouse CD8 anti-mouse CD11b), anti-Gr-1, and anti-mouse CD45.

6. 70-μm Cell strainer.

7. 3-mL Syringe.

8. 1% PFA in PBS.

9. 10 cm Cell culture petri dish.

10. 15 mL Conical tube.

3   Methods

3.1 Isolation of Mouse Primary Microglial Cells

1. Coat the 10 cm cell culture petri dish with 5 mL of 10 μg/mL PDL for 2 h at room temperature. Wash the dish with distilled water three times before use.

2. Collect the postnatal day 1 (P1) or day 2 (P2), decapitate the pups, and place the heads into a petri dish containing 10-mL cold culture medium. Remove the meninges carefully under a dissection microscope.

3. Transfer the brain tissues from the petri dish to a 15 mL tube. Use the transfer pipette to disassociate the brain tissues by pipetting up and down for several times.

4. Spin down the brain tissue at 800 g for 5 min. Add 10 mL of brain digestion medium containing 0.125% trypsin and 250 µg/mL DNase I to digest the brain tissue for 15 min at a 37 °C water bath. Swirl frequently.

5. Add 1–2 mL of FBS to terminate the digestion. Use the transfer pipette to pipette up and down to disassociate the cells. Filter the cells using 100-µm cell strainer. Spin down for 5 min at room temperature.

6. Aspirate the supernatant and resuspend the pellets with 5-mL prewarmed culture medium. Count the cells. Seed the cells into a petri dish at a density of 0.25×10^4 cells/mL (10 mL). Incubate the flask at 37 °C with 5% CO_2 and replace the medium every 5 days.

7. After 8–10 days of culture, the cells grow almost confluence. Digest the cells with 0.05% trypsin, and split the cells from one dish into two to three dishes. After another 8–10 days, the cells will grow 100% confluence again. It is important to wait the cells to grow 100% confluence for further mild trypsinization with custom-made trypsin solution with 1:4 ratio in DMEM-F12 medium (without FBS).

8. Digest the cells at 37 °C for 15–25 min. Check every 5 min after 15 min of digestion. The astrocytes layer would be digested like a "skin". Wash the attached cells with PBS, and add the 0.05% trypsin to detach the cells for 10 min. The cells are harvested as microglia for phenotypic analysis by flow cytometry (*see* **Note 1**) and used for further infection experiments.

3.2 Isolation of Mouse Primary Neurons

1. Euthanize heavily pregnant mice (embryonic (E)17.5-E19.5). Decapitate the fetuses, and transfer the heads into a petri dish containing HBSS. Take out the brains and transfer to a new petri dish containing HBSS.

2. After removing the olfactory bulb and cerebellum, dissect the cortex in a new petri dish containing HBSS. Remove meninges by gently pulling apart the membrane from the cortex, and transfer to a new petri dish containing HBSS.

3. Cut the cortex into small pieces into a 15 mL tube using a transfer pipette. Rinse three times with 10 mL of HBSS at

800 g for 5 min. Add 10 mL of cell culture medium, spin down, and remove the supernatant (the tube should be left with 2–3 mL of culture medium before adding trypsin).

4. Dissociate the brain tissue by adding 0.5 mL of $10\times$ trypsin. Incubate at 37 °C for 10 min. Stop the reaction by adding 1 mL of FBS. Ring two times with 10 mL HBSS at $800 \times g$ for 5 min.

5. Make a single-cell suspension by mixing the cells well using a transfer pipette. Spin down the cells. Remove the supernatant and resuspend the cells in 10 mL HBSS, and mix well.

6. Pass the cell suspension with a 70 μM cell strainer. Count the cells. Centrifuge the cells at 800 g for 7 min, remove the supernatant, and resuspend the cell pellet in 80-μl FACS buffer per 10^7 cells.

7. Add 20 μL Non-neuronal cell biotin-antibody cocktail per 10^7 cells. Incubate at 4 °C for 10 min. Wash cells with 2 mL FACS buffer. Add 20 μL of anti-biotin microbeads per 10^7 cells, and mix cells by pipetting. Incubate for 20 min at 4 °C. Add up to 1 mL of FACS buffer and proceed to the separation steps.

8. Place a LS column in the magnetic multistand separator, and rinse it with 3 mL of FACS buffer. Apply cell suspension to the column. Follow the manufacturer's instructions to collect the unlabeled cells (neuron cells), and wash three times with 3 mL of FACS buffer.

9. Count cells. Culture 0.5×10^6 per well in a 12-well PDL-coated plate as described above in Subheading 3.1, **step 1**.

10. Purity check: Take 0.6×10^6 cells, and split equally into three tubes, and centrifuge at 800 g for 5 min. Remove the supernatant, add 250 μL cytofix/cytoperm, and incubate at RT for 20 min. Wash cells with $1\times$ perm/wash buffer, and spin at 800 g for 5 min. Remove the supernatant, add 100 μL FACS buffer, and add 1 μL Fc blocker. Incubate at RT for 12 min. Add 50 μL of primary Ab: anti-NeuN Ab. (1:100 dilution), and incubate at 4 °C for 30 min. Wash once with FACS buffer. Add 50 μL anti-Rabbit IgG-PE (1:50 dilution), and incubate at 4 °C for 30 min. Wash cells two times. Fix the cells in 0.3 mL 0.5% PFA and analyze by flow cytometer to check the purity of cells (*see* **Note 2**).

3.3 WNV Infection in Primary Microglial Cells and Neurons

1. Infection of Microglial Cells.
   (i) Seed cells in a 96-well plate at 0.2×10^6 cells per well in 200 μL culture medium, and incubate at 37 °C for 24 h.
   (ii) Dilute WNV stock in cell culture medium, and add the virus at indicated multiplicity of infection (MOI) for 1 h. Wash cells twice with culture medium and incubate at 37 °C.

(iii) At days 1 and 4 postinfection, collect culture supernatant. Spin down the supernatant at 800 g for 5 min, and transfer the supernatant to a new tube.

(iv) Store the supernatants at −80 °C.

2. Infection of Primary Neurons.

(i) Change medium 6–10 h after the neurons become attached, carefully remove the DMEM medium, and add 0.5 mL of neurobasal medium.

(ii) After 4 days, replace the medium with 0.25 mL of fresh medium/well, and infect cells with WNV. Incubate cells at 37 °C for 2 h. Wash cells two times with 0.5 mL of DMEM medium.

(iii) At indicated time points postinfection, collect culture supernatant as described in Subheading 3.3.1. Store samples at −80 °C.

3.4 Virus Titration by FFA

1. On day 1, seed Vero cells at 1×10^5 cells/well in 500 µL of medium in a 24-well plate.

2. On day 2, prepare tenfold serial dilutions of each sample in PBSG. Add 100 µL/well of each virus dilution (in duplicate) after removing medium from Vero cells (not too dry). Incubate the plate at 37 °C for 1 h, and shake plate every 15 min. Add 1 mL/well of methylcellulose overlay (warm overlay to 37 °C in water bath or incubator) to the well. Incubate the plate in a 37 °C incubator for 36 h.

3. Remove the overlay, and wash cells three times with PBS (~0.6 mL/well), and allow cell monolayer to dry to increase cell adherence during IHC (5–10 min). Fix the cells with methanol:acetone (1:1) for at least 30 min at −20 °C.

4. Discard the methanol:acetone from each well, and allow plates to air-dry at room temperature for 30–60 min. Add 300 µL of blocking buffer and incubate 15–30 min at room temperature with platform rocking. Remove blocking solution, add 200 µL of rabbit anti-WNV polyclonal antibody (1:2000 dilution), and incubate at room temperature for 1 h.

5. Wash three times with PBS (0.5 mL/ well). Add 200 µL of goat anti-rabbit IgG HRP (1:2000 in blocking solution), and incubate at room temperature for 1 h.

6. Wash three times with PBS, and add 200 µL of AEC Developing solution as per manufacturer's instructions. Add three drops of each reagent (there are four solutions in the above kit) in 5 mL of PBS. Incubate 5–10 min at room temperature in the dark until spots are fully defined with minimal background. Stop the reaction by washing twice with H_2O. Let the plate dry and count the spots. Select a dilution with easily distinguished foci to calculate the virus titers (*see* **Note 3**).

3.5 Studying WNV-Induced Inflammation in Mouse Brains

1. Collection of brain tissues from WNV-infected mice: 5- to 8-week-old B6 mice are inoculated intraperitoneally (i.p.) with 100 FFU of WNV 385–99 and then transcardially perfused with 30 mL of cold PBS to deplete intravascular leukocytes at different time points postinfection. Mouse brains are placed in RMPI media to be used for isolation of brain leukocytes.

2. Study inflammatory infiltrates in the brain:

   (i)   Brain tissues are collected in 5 mL RPMI supplemented with 5% FBS and homogenized gently by pressing through a 70 μm cell strainer using the plunger of 3-mL syringe. Keep all the steps on ice.

   (ii) Harvest the cells in a 15-mL tube, and spin down at 800 g for 5 min at 4 °C. Resuspend the cell pellet in 7 mL of PBS with 2% FBS mixed with 3 mL of 90% Percoll in PBS.

   (iii) After 5 min (800 g) at 22 °C, the suspension is underlaid with 1 mL of 70% Percoll in RPMI and then centrifuged at (800 g) for 20 min at room temperature (no brake).

   (iv) The leukocytes at the interphase are isolated.

   (v)   Brain leukocytes are stained with antibodies to CD3, CD4, CD8, CD11b, Gr-1, and CD45; fixed with 1% PFA in PBS; and examined with a flow cytometer. Dead cells are excluded on the basis of forward and side light scatter (*see* **Note 4** for phenotypic description of brain leukocytes).

4   Notes

1. *Isolation of primary microglial cells.* The purity of the microglia is determined by flow cytometry quantification of CD11b$^+$ cells as 90–95%.

2. *Isolation of primary neurons.* The purity of the neurons is determined by flow cytometry quantification of NeuN$^+$ cells as 90–95%.

3. *WNV titration by FFA*: Count the spots in each well, choose the dilution shown clear spot counts of 5–100, and take the average count of the duplicate wells of the same dilution. Viral titers are calculated as focus-forming units per mL (FFU/mL) = the average count of spots/(dilution × volume of virus inoculum in each well). For example, if 20 and 24 spots are counted for duplicate wells of the 1×10^{-6} dilution, then WNV titer is calculated as 2.2×10^8 FFU/mL [22 (the average of the spot counts of the duplicate wells)/(10^{-6} (virus dilution) × 0.1 mL (virus inoculum)].

4. *Studying WNV-induced inflammation in the brain*: Phenotypic analysis of brain leukocytes reveals the following major infiltrating inflammatory cells in WNV-infected brains: (1) CD4$^+$ T cells (CD45$^+$CD3$^+$CD4$^+$); (2) CD8$^+$ T cells (CD45$^+$CD3$^+$CD8$^+$); (3) monocytes/macrophages (CD11b^{hi}CD45^{hi}); and (4) neutrophils (CD45$^+$Gr1$^+$). In addition, brain leukocytes include CNS residential cells, such as naïve microglial cells (CD11b^{lo}CD45^{lo}) and activated microglial cells (CD11b^{hi}CD45^{lo}) which can be differentiated from infiltrating monocytes/macrophages based on the cell surface expression levels of CD11b and CD45 [14, 22]. WNV-induced neurological diseases are caused by neuronal degeneration, a direct result of viral infection, and bystander damage from the inflammatory immune response induced by CNS residential cells and/or infiltrating immune cells [23, 24].

Acknowledgments

This work was supported in part by NIH grants R01 AI127744 (T.W.) and R21 AI140569 (T.W.).

References

1. Carson PJ, Konewko P, Wold KS et al (2006) Long-term clinical and neuropsychological outcomes of West Nile virus infection. Clin Infect Dis 43(6):723–730. https://doi.org/10.1086/506939

2. Ou AC, Ratard RC (2005) One-year sequelae in patients with West Nile virus encephalitis and meningitis in Louisiana. J La State Med Soc 157(1):42–46

3. Sadek JR, Pergam SA, Harrington JA et al (2010) Persistent neuropsychological impairment associated with West Nile virus infection. J Clin Exp Neuropsychol 32(1):81–87. https://doi.org/10.1080/13803390902881918

4. Nash D, Mostashari F, Fine A et al (2001) The outbreak of West Nile virus infection in the New York City area in 1999. N Engl J Med 344(24):1807–1814

5. Petersen LR, Brault AC, Nasci RS (2013) West Nile virus: review of the literature. JAMA 310(3):308–315. https://doi.org/10.1001/jama.2013.8042

6. Beasley DW, Li L, Suderman MT et al (2002) Mouse neuroinvasive phenotype of West Nile virus strains varies depending upon virus genotype. Virology 296(1):17–23. https://doi.org/10.1006/viro.2002.1372

7. Samuel MA, Diamond MS (2006) Pathogenesis of West Nile virus infection: a balance between virulence, innate and adaptive immunity, and viral evasion. J Virol 80(19):9349–9360. https://doi.org/10.1128/JVI.01122-06

8. Arjona A, Foellmer HG, Town T et al (2007) Abrogation of macrophage migration inhibitory factor decreases West Nile virus lethality by limiting viral neuroinvasion. J Clin Invest 117(10):3059–3066. https://doi.org/10.1172/JCI32218

9. Daniels BP, Holman DW, Cruz-Orengo L et al (2014) Viral pathogen-associated molecular patterns regulate blood-brain barrier integrity via competing innate cytokine signals. MBio 5(5):e01476–e01414. https://doi.org/10.1128/mBio.01476-14

10. Wang T, Town T, Alexopoulou L et al (2004) Toll-like receptor 3 mediates West Nile virus entry into the brain causing lethal encephalitis. Nat Med 10(12):1366–1373. https://doi.org/10.1038/nm1140

11. Zhang B, Patel J, Croyle M et al (2010) TNF-alpha-dependent regulation of CXCR3 expression modulates neuronal survival during West Nile virus encephalitis. J Neuroimmunol

224(1–2):28–38. https://doi.org/10.1016/j.jneuroim.2010.05.003

12. Quick ED, Leser JS, Clarke P et al (2014) Activation of intrinsic immune responses and microglial phagocytosis in an ex vivo spinal cord slice culture model of West Nile virus infection. J Virol 88(22):13005–13014. https://doi.org/10.1128/jvi.01994-14

13. Shrestha B, Gottlieb D, Diamond MS (2003) Infection and injury of neurons by West Nile encephalitis virus. J Virol 77(24): 13203–13213. https://doi.org/10.1128/jvi.77.24.13203-13213.2003

14. Luo H, Winkelmann ER, Zhu S et al (2018) Peli1 facilitates virus replication and promotes neuroinflammation during West Nile virus infection. J Clin Invest 128(11):4980–4991. https://doi.org/10.1172/JCI99902

15. Parquet MC, Kumatori A, Hasebe F et al (2001) West Nile virus-induced bax-dependent apoptosis. FEBS Lett 500(1–2):17–24. https://doi.org/10.1016/s0014-5793(01)02573-x

16. Brehin AC, Mouries J, Frenkiel MP et al (2008) Dynamics of immune cell recruitment during West Nile encephalitis and identification of a new CD19+B220-BST-2+ leukocyte population. J Immunol 180(10):6760–6767. https://doi.org/10.4049/jimmunol.180.10.6760

17. Glass WG, Lim JK, Cholera R et al (2005) Chemokine receptor CCR5 promotes leukocyte trafficking to the brain and survival in West Nile virus infection. J Exp Med 202(8): 1087–1098. https://doi.org/10.1084/jem.20042530

18. Lim JK, Obara CJ, Rivollier A et al (2011) Chemokine receptor Ccr2 is critical for monocyte accumulation and survival in West Nile virus encephalitis. J Immunol 186(1): 471–478. https://doi.org/10.4049/jimmunol.1003003

19. Sitati E, McCandless EE, Klein RS et al (2007) CD40-CD40 ligand interactions promote trafficking of CD8+ T cells into the brain and protection against West Nile virus encephalitis. J Virol 81(18):9801–9811. https://doi.org/10.1128/JVI.00941-07

20. Sitati EM, Diamond MS (2006) CD4+ T-cell responses are required for clearance of West Nile virus from the central nervous system. J Virol 80(24):12060–12069. https://doi.org/10.1128/JVI.01650-06

21. Xiao SY, Guzman H, Zhang H et al (2001) West Nile virus infection in the golden hamster (Mesocricetus auratus): a model for West Nile encephalitis. Emerg Infect Dis 7(4):714–721

22. Welte T, Aronson J, Gong B et al (2011) Vgamma4+ T cells regulate host immune response to West Nile virus infection. FEMS Immunol Med Microbiol 63(2):183–192. https://doi.org/10.1111/j.1574-695X.2011.00840.x

23. Maximova OA, Pletnev AG (2018) Flaviviruses and the central nervous system: revisiting neuropathological concepts. Annu Rev Virol 5(1): 255–272. https://doi.org/10.1146/annurev-virology-092917-043439

24. Ghosh Roy S, Sadigh B, Datan E et al (2014) Regulation of cell survival and death during Flavivirus infections. World J Biol Chem 5(2): 93–105. https://doi.org/10.4331/wjbc.v5.i2.93

Chapter 7

Detection of West Nile Virus Envelope Protein in Brain Tissue with an Immunohistochemical Assay

Kathleen T. Yee and Douglas E. Vetter

Abstract

Immunohistochemistry is a valuable tool for probing not only scientific questions but also clinical diagnoses. It provides power from localization of a protein within the milieu of a tissue section that may reflect positioning within or beyond the boundaries of a cell that is representative of the tissue at a discrete moment in time. The method can be applied broadly, including to tissues under normal, developmental, chemically, or genetically altered conditions and disease states.

Disease manifesting from West Nile virus infection ranges from acute, systemic febrile symptoms to compromise of central nervous system function. Immunohistochemistry has been used to assess WNV infection in the nervous system in postmortem and experimental conditions, despite the lack of understanding of the precise route of viral entry. In addition to imprecise knowledge of initial viral entry into cells and whether entry is even the same between cell types, the fact that spontaneous viral mutations and environmental pressures from climate change may alter the prevalence of the disease state across geographical and climatological boundaries highlights the need for continued assessment of infection. Immunohistochemistry is a useful way to assess these aspects of WNV infection with the aim being to better understand the organs and cell types that are compromised by WNV infection. This chapter outlines how this can be carried out on brain tissue, but the procedures discussed can also be applied more broadly on tissue outside of the central nervous system.

Key words Antibody staining, WNV, Fixed brain tissue

1 Introduction

Immunohistochemistry is an approach that melds microscopic anatomy and components of immune biology. It exploits the use of antibodies (produced under normal conditions by organisms in response to foreign antigens) to bind experimenter-chosen protein epitopes to probe tissue for basic science investigations [1–6] and clinical diagnostics (e.g., [7]). Unlike blot techniques such as Western (immune) blotting, immunohistochemistry allows investigators to anatomically localize proteins of interest.

Fengwei Bai (ed.), *West Nile Virus: Methods and Protocols*,
Methods in Molecular Biology, vol. 2585, https://doi.org/10.1007/978-1-0716-2760-0_7,
© The Author(s), under exclusive license to Springer Science+Business Media, LLC, part of Springer Nature 2023

Historically, immunohistochemistry has been possible due to the initial identification of what would later become known as antibodies; the "anti-toxins," as they were called at the time, were successful treatment for diphtheria and tetanus [8]. Since that time, there has been an increased understanding of antibody-antigen interactions that has allowed for an explosive evolution of tools for research and medical application. The recognition that labeling of antibodies would allow for the detection of their protein partner began with antibody conjugation to visible tags [9, 10], expanded to fluorescent tags [11], and tags for immunoenzyme techniques [12] and what is in common use today—more durable fluorophore tags that are visualized with confocal microscopy (*see* review [13]).

Through the application of immunohistochemistry, the precise extracellular, intracellular, or subcellular localization of a protein of interest within a tissue section can be determined 'in situ.' Successful immunohistochemical labeling can be achieved with slight variations in the immunohistochemical technique which include trade-offs between the length of procedure and numbers of different steps, while other variations lead to increasing levels of amplification. These approaches include the peroxidase-anti-peroxidase (PAP)-based chromogenic method, useful, for example, in labeling neurons expressing subsets of neurotransmitters or their synthetic enzymes [3, 14, 15], the related but now more common avidin-biotin complex (ABC) protocol [16] that has essentially replaced the practice of PAP procedures, and, for example, has been used to immunolabel cochlear neuronal fibers [5] and the indirect method utilizing a tagged secondary antibody and has been useful for visualizing cochlear synapses as one example [4, 5, 17]. The indirect immunohistochemical method is typically used to label immunopositive structures using diaminobenzidine (DAB) as a chromogen (although other chromogens are available) and hydrogen peroxide to catalyze the deposition of the chromogen via the interaction of H_2O_2 with the peroxide of the PAP complex or the ABC [6]. Further, post-DAB reaction intensification can be accomplished by depositing heavy metals at the site of chromogen accumulation, typically by use of nickel with or without cobalt [18]. Immunolocalization using either DAB or fluorescent tags is useful for visualizing adult neuronal structures [4, 5], cells during embryonic development in whole mount preparations [2] and revealing viral loci and subsequent consequences due to Zika virus infection [1]. Related to the latter, immunohistochemical staining can be applied to assess viral infection in the central nervous system that may contribute to encephalitis caused by West Nile virus. While the advances in fluorescence microscopy, including the now commonplace confocal microscopy, but also the increasing availability of super-resolution microscopy has driven the immunolabeling procedures toward the use of fluorescently tagged secondary antibodies, the DAB technique still has its uses, especially when extreme amplification is required. Building on the ABC technique, amplification via iterative application of biotinylated tyramine (BT) can produce a large amplification locus that acts as a reservoir for binding extreme numbers of peroxidase molecules, each of which can catalyze the deposition of DAB chromogen. Reaction with DAB in the BT amplification procedure yields orders of magnitude greater sensitivity than the standard DAB procedures and can be used for visualizing proteins at very low expression levels. As an example, an antibody to a relatively normally expressed protein (substance P) was visualized with primary antibody dilutions of 1: 1,000,000 [19]. In this case, diluting the primary antibody to such an extreme serves as a proxy for conditions in which immunolabel of very few molecules (such as in the case of low-abundance proteins) are labeled with standard primary antibody dilutions.

West Nile virus (WNV), first identified in a febrile 37-year-old female in the West Nile district of Uganda in 1937 [20], is a member of the *Flaviviridae* family that includes dengue and Zika viruses. In the intervening 80 years, WNV has led to infection in six continents—Africa, Europe, Asia, Australia and the Americas, both North and South (*see* review [21]). WNV transmission persists in a life cycle between arthropod and vertebrate hosts, and mosquitoes and birds are the principal players in the WNV life cycle. Birds are the amplifying hosts as they develop transient high levels of viremia postinfection; during this period, bites from any mosquitoes have a high likelihood of leading to mosquito infection. Twenty-five mammalian species plus humans and horses have demonstrated infectability by WNV and a risk for disease [22, 23]. The most common mode of WNV infection for humans is through the bite of an infected mosquito, but other routes for infection have been documented via blood transfusion [24, 25], organ transplant [26], transplacental transfer [27], and likely exposure to infected breast milk [25, 28].

More than 60 species of mosquitoes can be infected with WNV [29]. The primary vector for WNV is the *Culex* mosquito, and *Culex* subgroups populate different geographic niches [30–35]. Local populations of mosquito species must be high, and the presence of WNV in a particular mosquito species is necessary to maintain the transmission cycle from bird to mosquito back to bird again. The preference of mosquitoes may shift to bridging vectors such as humans and domesticated animals. While *Culex* mosquitoes prefer avian hosts, they will feed on mammals [36, 37]. Additionally, it is possible for WNV to overwinter in infected female mosquitos (females being the predominant of the species to engage in mosquito bite blood feasts) and pass the virus on to their larvae in the next season [38].

WNV has been found to infect more than 300 bird species in North America [29]. Surveillance of dead birds between August and December 1999 during the New York City outbreak showed that nearly 90% of the West Nile-infected dead birds were crows [39, 40]. However, a study examining avian vectors found that the northern cardinal and house sparrows are the most important amplifying hosts for WNV in St. Tammany Parish, Louisiana [41]. A similar study in the vicinity of Washington, D.C. showed that robins are the primary WNV amplification host, while the more populous house sparrows contributed only in a minor way and were avoided by mosquitos [42]. These studies are important and suggest that WNV-amplifying hosts vary by region; this would not be surprising since there are seasonal shifts in host preference from birds to mammals by the most common mosquito vector, *Culex tarsalis*, along the California coast [43] and Colorado [44]. In the northeast of the continental USA, *Culex pipiens* shifts its blood meal preference from June to September from American robins to catbirds and mourning doves [45]. *Culex pipiens* in the Washington, D.C. vicinity also reportedly shift their preference during the year from American robins to mammals, including humans [42].

It has been recognized that WNV infection leads to varied symptoms. Most individuals contract acute, systemic febrile disease or West Nile fever (WNF), and a minority of individuals develop disease that is neuroinvasive including aseptic meningitis (West Nile meningitis (WNM)), encephalitis (West Nile encephalitis (WNE)), or West Nile virus-associated acute flaccid myelitis (WN-associated AFM) [46]. These varied symptoms manifested in different parts of the world. Epidemics of WN encephalitis were observed in some African countries, eastern Europe, the Middle East [46, 47], North America [48, 49], and southeastern Romania [50], while WNV infections in central and southern Africa predominantly resulted in systemic illness and accompanying fever but lacked involvement of the central nervous system (CNS) [51]. Serological and molecular biological analyses revealed that different viral strains led to the WNV infections seen around the world [52–54]. Lanciotti and colleagues report that there are two main lineages of WNV (lineage 1 and lineage 2) [55]. Lineage 1 was found in eastern Europe, some African countries, the Middle East, and North America and is the lineage with a greater propensity to cause WN encephalitis. In contrast, lineage 2 brought about only systemic illness and fever and an absence of encephalitis. It has been speculated that WNV in southern and central Africa was endemic and that the population may have had background immunity during that time. However, in the 1990s, lineage 2 WNV began to spread to the Mediterranean and Europe, paving the way for an outbreak in Greece in 2010 [56]. Unexpectedly, the spread of the WNV lineage 2 during the 2010 WNV outbreak in Greece [57] nonetheless led to WN encephalitis in many infected individuals [56, 58] despite lineage 2 historically being much more innocuous [59].

The consequence of WNV infection as a neuroinvasive disease has prompted studies that have examined nervous system infectivity by WNV. Examination of human neuropathology post-WNV infection has revealed that the spinal cord is more heavily infected by WNV than the brain [60]. However, a lack of knowledge of the precise time between infection and death for analysis of postmortem tissues presents limitations to interpretations. Using nonhuman mammalian animal models eliminates these unknowns. Many open questions remain as to how WNV enters the nervous system (*see* review [61]); precise localization of infected WNV nervous system cells following known timings postinfection can contribute to answering these questions. Additionally, the evolving nature of WNV combined with environmental pressures from climate change may lead to genetic changes in WNV (*see* review [21]) that could alter its patterns of infectability and that can be studied by immunohistochemical localization of WNV in the nervous system.

Immunohistochemistry is a straightforward procedure that can be used to track WNV infection of cells under experimental conditions (Figs. 1 and 2). There are issues that persons carrying out the technique should be aware of that are described in the following protocol. Additional caveats and alternatives are provided in the Notes section.

2   Materials

Note: Follow all proper protocols for disposal of any waste materials.

2.1   General Solutions, Equipment, and Supplies

1. Deionized water: Use deionized water to prepare all solutions.

2. 10× Phosphate buffered saline (10× PBS) stock solution: 0.072 M Na_2HPO_4, 0.022 M $NaH_2PO4{\bullet}H_2O$, 1.266 M NaCl.

3. 0.4 M Phosphate buffer stock solution: 0.065 M $NaH_2PO4{\bullet}H_2O$, 0.33 M Na_2HPO_4.

4. 1× Phosphate buffered saline (1× PBS) working solution: Dilute 1 part of 10× PBS with 9 parts of deionized water.

5. 0.1 M Phosphate buffer working solution: Dilute 1 part of 0.4 M phosphate buffer with 3 parts of deionized water.

6. Orbital Shaker (horizontal, no incline angle).

7. Alcohol-soluble markers.

2.2   Solutions for Perfusion

1. 10% Paraformaldehyde (stock solution): 10% paraformaldehyde in deionized water. The paraformaldehyde is combined with deionized water and warmed to a maximum of 60 °C. Add paraformaldehyde to the warm water, and stir vigorously with a

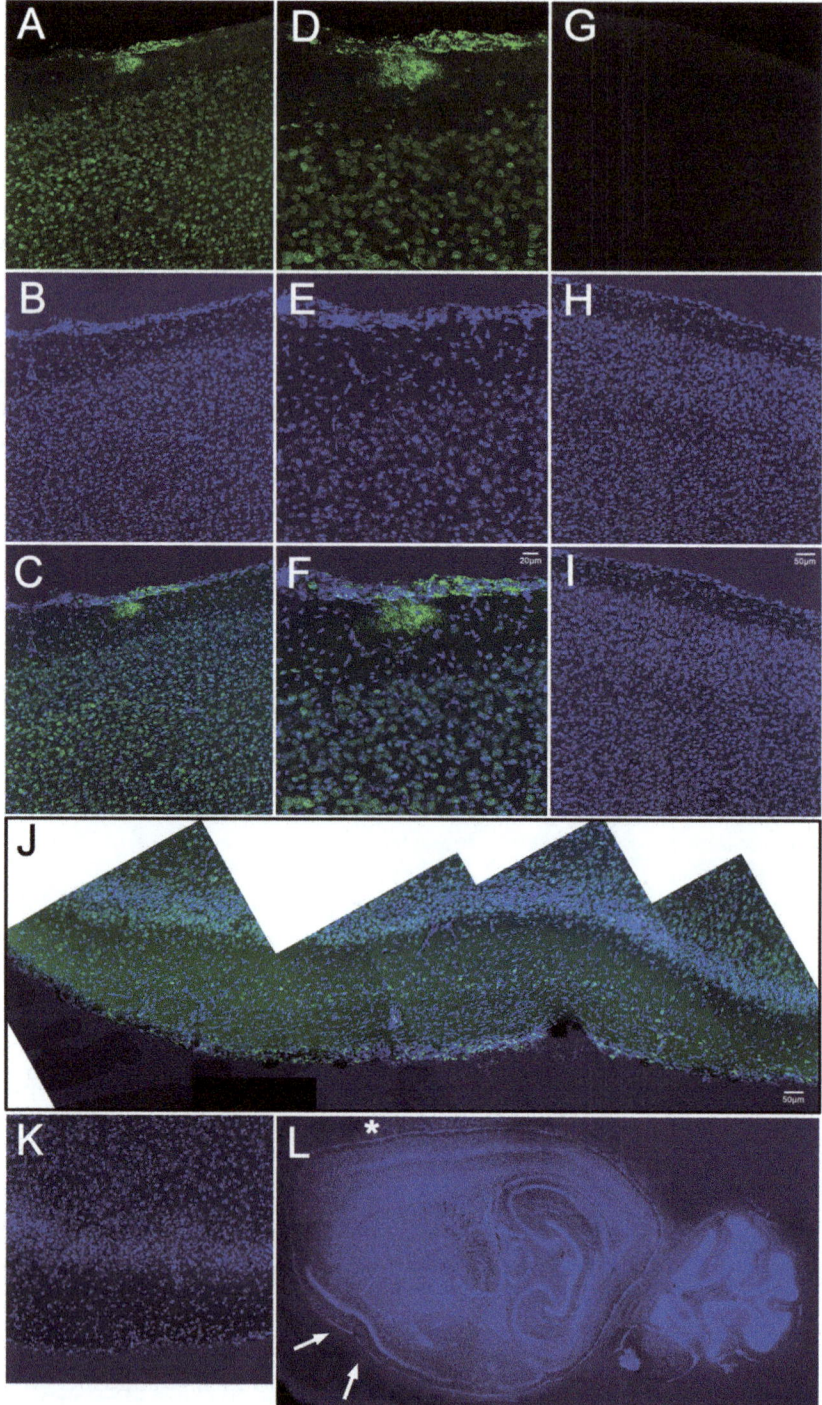

Fig. 1 West Nile virus envelope protein localization in cells of the dorsal frontolateral cortex and piriform cortex of IFNRA1-/- mice 6 days after infection with 2000 PFU of WNV (strain CT2741) in 100 μl of PBS by i.p. injection (**a–c**, **d–f**, **j**). Dorsal frontolateral cortex, indicated by the asterisk, and a portion of piriform cortex, indicated by the two arrows show localization of West Nile virus envelope protein (**l**). High levels of West Nile virus envelope protein are expressed on the dorsal frontolateral surface with staining extending just below the pial surface as well as ventrally in layers 2 and 3 (**a**). A higher magnification of **a** is shown in **d**. DAPI stains of the sections in **a** and **d** are shown in **b** and **e**, respectively. The merged views are shown in **c** and **f**.

stir bar. After the paraformaldehyde is in suspension, add 10 N NaOH dropwise until the solution is almost clear. The paraformaldehyde should then be filtered immediately with coarse filter paper into a bottle or other container sitting in an ice bath. The immediate cooling is required to avoid excess formation of formic acid in the solution. Store the 10% paraformaldehyde at 4 °C.

2. 4% Paraformaldehyde (working solution): 4% paraformaldehyde in 0.1 M phosphate buffer. Combine 40 mL of 10% paraformaldehyde stock solution with 25 mL of 0.4 M phosphate buffer, and add deionized water to a total volume of 100 mL.

2.3 Solutions for Cryoprotection

1. 20% Sucrose: 20% sucrose in 0.1 M phosphate buffer.
2. 30% Sucrose: 30% sucrose in 0.1 M phosphate buffer.

2.4 Supplies for Tissue Sectioning

1. Sliding microtome.
2. Sliding microtome stage.
3. Sliding microtome blade.
4. Circular bubble level.
5. Cryoprotected brain tissue.
6. Tissue forceps.
7. M1 embedding matrix.
8. Dry ice.
9. 24-Well untreated tissue culture plates.
10. Paint brush, medium fine.
11. 0.1 M phosphate buffer.

2.5 Solutions and Supplies for Immunohistochemical Staining of Free-Floating Brain Sections

1. Flat bottom airtight container.
2. Plastic petri dishes.
3. Plastic agglutination tray (may substitute 12- or 24-well plate).
4. Plastic agglutination tray with holes (*see* Preparation of Supplies).
5. Glass hooks made from 6″ Pasteur pipettes.
6. Normal donkey serum.
7. 10% Triton X-100 in 1× PBS.

Fig. 1 (continued) West Nile virus envelope protein is also expressed by cells in the piriform cortex (**j**). An absence of WNV envelope protein staining correlates with no WNV infection in dorsal frontolateral cortex (**g** and **i**; DAPI stain visualized alone (**h**)) or in the merged views for dorsal frontolateral cortex (**i**) and piriform cortex (**k**) in uninfected control mice. 50 μm scale bar shown in **i** and **j** applies to **a**–**c** and **g**–**k**. 20 μm scale bar in **f** applies to **d**–**f**

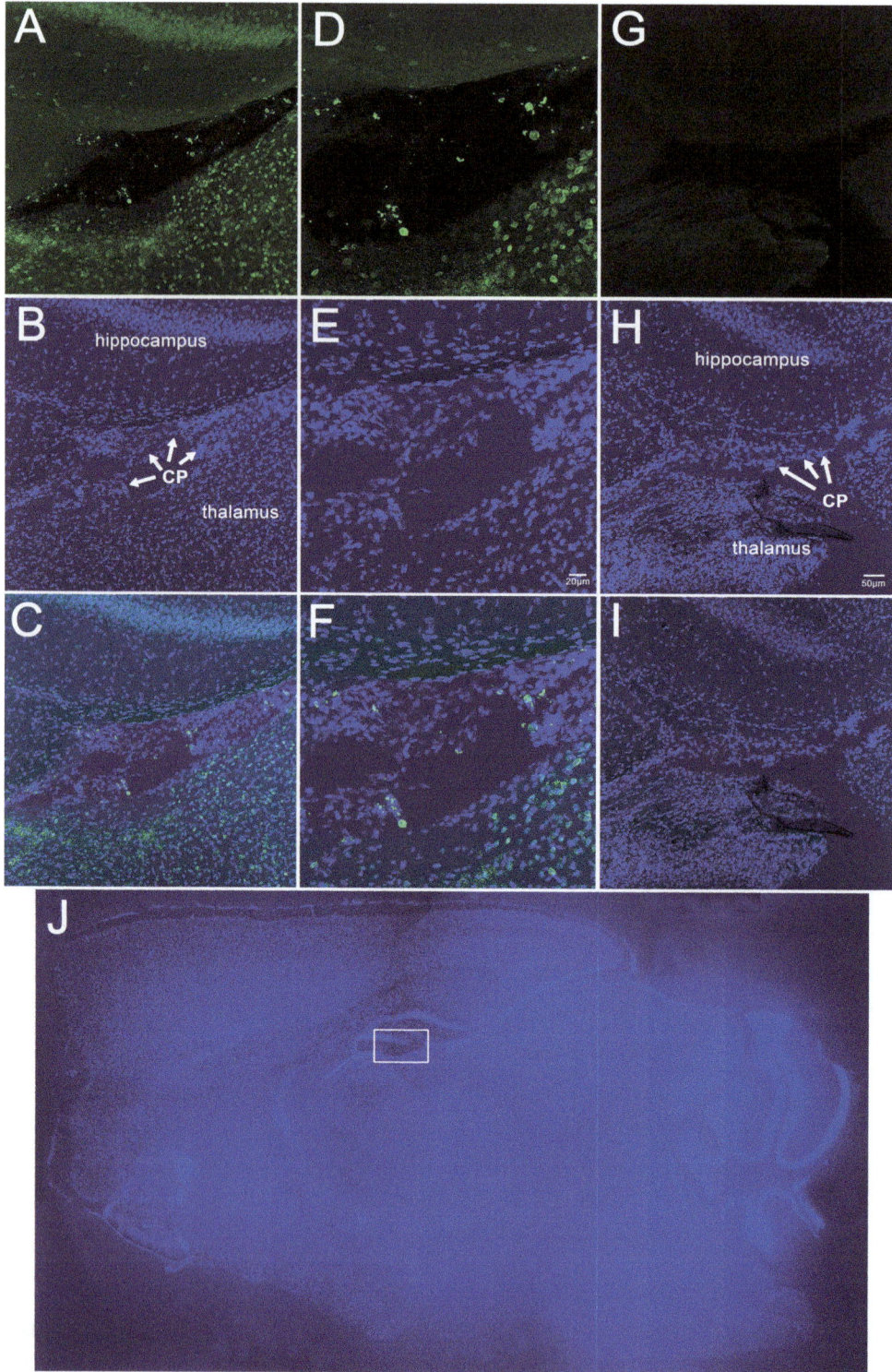

Fig. 2 West Nile virus envelope protein localization in cells of the choroid plexus, hippocampus, and thalamus of IFNRA1-/- mice 6 days after infection with 2000 PFU of WNV (strain CT2741) in 100 μl of PBS by i.p. injection (**a–c**, **d–f**, **j**). The region shown at high magnification (**a–f**) is indicated by the boxed area in **j**. High levels of West Nile virus envelope protein are expressed in the choroid plexus (CP, **a**; higher magnification

8. Blocking solution: 5% normal donkey serum and 0.5% Triton X-100 in 1× PBS.

9. Polyclonal antibody against West Nile virus envelope protein.

10. Secondary antibody, goat anti-rabbit (H + L) Alexa 488.

11. Glass hooks made from 6″ Pasteur pipettes.

12. Bunsen burner.

13. 2000-mL beaker.

14. Paper towel.

15. Stir bar.

16. Slide racks.

17. Deep rectangular glass dish (large enough to accommodate 1–2 slide racks).

18. Covered plastic container (large enough to accommodate all of the slide racks).

19. Gelatin (270 bloom).

20. 1% double gelatin-coated 25 × 75 mm glass slides: Insert 25 × 75 mm glass slides into the slide racks. Add detergent to the container followed by tap water, and allow suds to form. Cover container and allow to soak overnight. The following day, rinse thoroughly with tap water, and allow to soak in two changes of tap water (30 min each) and two changes of deionized water (30 min each). During the final soak in deionized water, prepare the gelatin-coating solution.

21. 1% gelatin for slide coating: To 500 mL of deionized water in a 2000-mL beaker, gradually sprinkle 5 gm of gelatin on the surface of the water, allowing gelatin to swell before adding more gelatin. Microwave the deionized water and gelatin until the gelatin dissolves. Add 0.5 gm chromium potassium sulfate and a stir bar, and stir until the chromium potassium sulfate dissolves. Maintain the temperature of the gelatin solution between 35 and 45°C for slide coating. For the first gelatin coat, immerse slides in 1% gelatin for 5 min, drain off gelatin, and allow to dry. For the second gelatin coat, immerse the glass slides in 1% gelatin for 10 min, drain, and set aside to dry. When the gelatin is dry, remove slides from the racks, and store for future use.22. Paint brushes, fine (0/5–0/10).

Fig. 2 (continued) of **a** in **d**). West Nile virus envelope protein is also expressed by cells in the hippocampus and thalamus (**a** and **d**). DAPI stain of the sections in **a** and **d** are shown in **b** and **e**, respectively. The merged views are shown in **c** and **f**. No WNV infection correlates with a lack of WNV envelope protein staining in the choroid plexus, hippocampus, and thalamus (**g** and **i**); the same region stained only for DAPI is shown in **h**. 50 μm scale bar shown in **h** applies to **a–c** and **g–i**. 20 μm scale bar in **e** applies to **d–f**

2.6 Supplies for Counterstaining and Coverslipping Brain Sections

1. 4′,6-diamidino-2-phenylindole [DAPI] (stock solution 10 mg/mL).

2. Glass coverslips.

3. Blunted hypodermic needle.

4. TES fluorogel (or other coverslipping medium containing anti-fade reagent).

2.7 Equipment for Viewing Fluorescent-Stained Specimens

1. Epifluorescence microscope or confocal microscope.

2. Fluorescent filter cubes or confocal lasers set to visualize fluorescence stained specimens.

3 Methods

3.1 Perfusion

1. Transcardially perfuse mice first with room temperature 1× phosphate buffered saline. Perfuse this rinse solution until blood ceases to exit the heart (approximately 10 mL) (*see* **Note 1**).

2. Then transcardially perfuse mice with room temperature 4% paraformaldehyde. 10−20 mL of 4% paraformaldehyde is sufficient to perfuse the entire body of an adult mouse (*see* **Note 1**).

3. Remove the brain (or other desired tissue) from the body and place tissue in a scintillation vial containing ~5 mL of 4% paraformaldehyde; place on an orbital shaker with gentle agitation at low RPM, and post-fix for 1 h at room temperature (*see* **Note 2**).

3.2 Cryoprotection

1. Following the 1-h post-fix period, decant the 4% paraformaldehyde, add ~5 mL 20% sucrose, and place on an orbital shaker with gentle agitation at low RPM at room temperature. If leaving overnight, place on an orbital shaker at 4°C (*see* **Note 3**).

2. When tissue has sunk to the bottom of the vial containing 20% sucrose, decant the 20% sucrose and add ~5 mL 30% sucrose. Place on an orbital shaker at room temperature or orbital shaker at 4°C if leaving overnight (*see* **Note 3**).

3. When tissue has sunk to the bottom of the vial containing 30% sucrose, the tissue has been infiltrated by the sucrose solution and may be progressed to the next step.

3.3 Tissue Sectioning by Sliding Microtome

1. Fill each well of a 24-well plate ~half-full with 0.1 M phosphate buffer (*see* **Note 4**).

2. Label the lid of the 24-well plate with pertinent information regarding the tissue specimen with an alcohol soluble marker (*see* **Note 5**).

3. Secure the microtome blade and sliding microtome stage into their respective positions (*see* **Note 6**).

4. Adjust the sliding microtome stage, as needed, with the stage leveling screws so that the X and Y axes are horizontal as assessed by a circular bubble level (*see* **Note 7**).

5. Add a 4 × 4 cm area of M1 embedding matrix to the stage (*see* **Note 8**).

6. Add dry ice to the stage to initiate the freezing process.

7. As the M1 embedding matrix begins to freeze, beginning at the edges, its appearance will change from clear to white.

8. Place the brain tissue on unfrozen M1 embedding matrix with tissue forceps.

9. Position/freeze tissue into place at the appropriate angle for sectioning, or use the stage leveling screws to adjust the angle of the stage and secured tissue (*see* **Note 9**).

10. Set the section thickness to 40 μm.

11. After each 40 μm section is cut, use the paint brush to move the section from the microtome blade to the first well in the 24-well plate.

12. Place the next cut section into the adjacent well and so on.

13. Periodically check that there is always dry ice in contact with the microtome stage so that the microtome stage and adhered tissue remain frozen.

14. Place the 25th section in the same well as section 1 and so on.

15. Continue sectioning until the tissue has been completely cut through the region of interest.

3.4 Immuno-histochemical Staining of Floating Brain Sections

1. Preparation of Supplies.

1.1. Plastic agglutination tray with holes: created by making holes in the plastic agglutination tray(s) with a 22-gauge needle—punch holes from the inside to the outside of the well (*see* **Note 10**).

1.2. Glass hooks made from 6″ Pasteur pipettes: Seal off the small end by melting it using a Bunsen burner. Then shift the end of the Pasteur pipette so that the region ~0.5 cm from the end of the Pasteur pipette is positioned over the hottest part of the flame. When this area melts and bends to create a hook, quickly remove the Pasteur pipette away from the flame. Glass hooks can be washed and reused.

1.3. 4′,6-diamidino-2-phenylindole (DAPI) working stock: 1 μl of DAPI stock solution in 10 mL of 0.9% saline.

2. Immunohistochemical Staining.

2.1. Day 1.

2.1.1. Move the desired number of sections from the 24-well collection plate using a glass hook into 250 μl per well of blocking solution in a plastic agglutination tray with no holes. Place the sections in the plastic agglutination tray into a humidity chamber (a flat-bottomed airtight container lined with a deionized water-saturated paper towel) for 30–60 min on a horizontal orbital shaker with gentle agitation (low RPM) at room temperature. Scale the volume of solution accordingly, to cover the tissue, if larger wells are used (*see* **Note 11**).

2.1.2. Prepare the antibody dilution (1% normal donkey serum and 0.1% Triton X-100 and Invitrogen West Nile virus envelope polyclonal antibody in 1× PBS) and the control vehicle solution lacking primary antibody, and dispense them into clean, labeled (an alcohol-soluble marker can be used) agglutination trays (250 μl per well).

2.1.3. Using glass hooks that are kept separate for use with or without primary antibody, move the sections from the blocking solution into either the wells containing primary antibody or the vehicle control solution. There is no wash step between the blocking solution incubation, and transfer to the primary antibody solution.

2.1.4. Place the agglutination tray(s) containing the brain sections back into the humidity chamber.

2.1.5. Place the humidity chamber on an orbital shaker with gentle agitation (low RPM), and incubate overnight at room temperature.

2.2. Day 2.

2.2.1. Using dedicated glass hooks for the sections incubated in primary antibody and vehicle control solution, transfer sections into a petri dish containing an agglutination tray with holes and 1× PBS for primary antibody incubated sections and a second petri dish containing an agglutination tray with holes tray and 1× PBS for the vehicle control sections, respectively.

2.2.2. Wash three times with 1× PBS: first wash 5 min, second wash 10 min, and third wash 15 min (*see* **Note 12**).

2.2.3. During the washes, prepare the secondary antibody dilution (1% normal donkey serum and 0.1% Triton X-100 and goat and rabbit (H + L) Alexa 488 in 1× PBS); all sections (primary antibody incubated and vehicle control) will be incubated in the secondary antibody. Protect this assembled solution from light (*see* **Note 13**).

2.2.4. Aliquot 250 μl of secondary antibody into the matching number of wells in an agglutination tray.

2.2.5. A single glass hook may be used to transfer the sections into the same corresponding number of wells containing secondary antibody.

2.2.6. Place the agglutination tray(s) containing the brain sections back into the humidity chamber.

2.2.7. Place the humidity chamber on the orbital shaker at low speed, and incubate for 1 h at room temperature. Protect the humidity chamber from light.

2.2.8. After the 1-h incubation, a single glass hook may be used to transfer the sections to an agglutination tray with holes inside of a petri dish containing 1× PBS.

2.2.9. Wash three times with 1× PBS: first wash 5 min, second wash 10 min, and third wash 15 min.

3.5 Mounting of Immunohistochemically Stained Tissue Sections Onto Glass Slides

1. Fill a petri dish with 1× PBS and 20 μl of 1% Triton X-100 in 1× PBS. Grasp the end of a 1% double-gelatin-coated slide designated for labeling, place the opposite short end of the slide into the petri dish, and use a paint brush to guide/mount an immunohistochemically stained brain section onto this 1% double-gelatin-coated slide. Mount the next closest adjacent section close to the previously mounted section. Repeat until all brain sections are mounted onto the slide(s) (*see* **Note 14**).

2. Allow the sections to air dry.

3.6 Counterstain for Nuclei with 4′,6-Diamidino-2-Phenylindole (DAPI)

1. Lay dry slides to be counterstained horizontally on the benchtop.

2. Cover all of the dry sections with 0.9% saline.

3. Decant the 0.9% saline from the slide.

4. Place the slides on a layer of Parafilm, apply the DAPI working stock to the slide to cover all of the brain sections on each slide, and incubate for a minimum of 2 min (*see* **Note 15**).

5. Rinse all of the brain sections completely three times with 1× PBS.

6. Perform three washes with $1\times$ PBS a minimum of 2 min each. Decant after each wash except the last wash.

7. Decant the $1\times$ PBS from one slide and apply 5–6 drops of TES fluorogel evenly spaced across the slide when using a full-size coverslip. Decrease the number of drops of TES fluorogel accordingly, when using smaller coverslips. Ensure that the TES fluorogel is spread evenly over all of the sections forming a single contiguous layer of solution (*see* **Note 16**).

8. Gently and slowly lower a cover glass using a blunted hypodermic needle onto the TES fluorogel without introducing any air bubbles (*see* **Note 16**).

9. Repeat from **step 7** until all slides are coverslipped.

3.7 View Immuno-histochemically Stained Sections

1. View with the appropriate filter cubes or lasers to visualize the secondary antibody tagged with Alexa Fluor 488 and 4′,6-diamidino-2-phenylindole (DAPI), i.e., fluorescein and ultraviolet filter cubes or confocal lasers set to excite the proper wavelengths (488 nm and 358 nm, respectively) (*see* **Note 17**).

4   Notes

1. The volume of perfusates can be decreased by clamping the descending aorta, provided that this will perfuse the tissues of interest; if not, perfuse the entire body. Ensure that the perfusion of $1\times$ phosphate buffered saline runs clear before switching to the 4% paraformaldehyde perfusate.

2. Use care when removing the brain and other tissues from the body to avoid damage to tissues or organs.

3. Avoid making sucrose solutions in $1\times$ PBS since the presence of saline and water leads to oxidation of the microtome blade.

4. Avoid using $1\times$ PBS as the tissue section collection buffer since as stated above; the presence of saline and water leads to oxidation of the microtome blade. Use 0.1 M phosphate buffer instead.

5. Any labeling done with an alcohol-soluble marker can be removed with 100% ethanol and the 24-well plates can be reused.

6. Ensure that the microtome blade and microtome stage are secured in a finger-tight manner. An improperly assembled microtome with loose components may compromise uniformity of section thickness or prevent sectioning altogether and can even lead to operator injury if the knife falls out.

7. Horizontal leveling of the stage provides maximal unimpeded sectioning distance when the tissue block being cut approaches the base of the specimen.

8. 30% sucrose or optimal cutting temperature (OCT) compound may be substituted for M1 embedding matrix. Apply any of these in a 4 × 4 cm area to a microtome stage at room temperature (not frozen). Application to a frozen stage will result in rapid freezing without securely adhering to the stage.

9. Before the M1 embedding matrix freezes around the base of the tissue, an additional small amount of M1 embedding matrix may be added around the base of the tissue for extra adhesion.

10. When making an agglutination tray containing holes, avoid punching holes from the outside to the inside; potential plastic edges extending from the edge of each hole inside of the well can cut and damage the brain sections (may substitute Netwells, Corning).

11. When transferring sections from one location to another using a glass rod, avoid sandwiching a brain section between a glass rod and any container surface; this can cause damage to the section(s).

12. A total wash time of 30 min (5 + 10 + 15 min washes) can be decreased by increasing the number of washes and decreasing their duration.

13. Ensure that all reagents are added to the secondary antibody solution, including the fluorescent fluorophores. Omission of the fluorescent fluorophores will lead to an absence of fluorescent signal.

14. The addition of a small amount of Triton X-100 to the mounting buffer (1× PBS) breaks the surface tension allowing brain sections to be more easily removed from the mounting buffer. It is critical to mount brain sections on to glass slides that have been treated to provide optimal adhesion. Charge-coated slides will not be strong enough to adhere many 40-μm-thick sections; gelatin-coated slides work well. It is logistically easier to mount tissue oriented in columns on a slide. Once sections have been mounted in one column, move one section's width to the right, and begin to mount the sections in the adjacent column and so on.

15. Using Parafilm as an underlayment provides a waterproof surface that can catch any DAPI spillage and provides an easy means for containment and hazardous waste disposal.

16. Air bubbles located between the specimen and coverslip may impede visualization of the specimen. Several steps can be taken to minimize trapping of air bubbles. Any areas of the slide uncovered but surrounded by TES fluorogel are prone to air bubble formation and more likely to occur if the coverslip is quickly dropped down onto the mounting medium. First,

ensure that the TES fluorogel forms a single continuous layer across the surface of the slide; this will reduce the likelihood that air will be trapped in the TES fluorogel-free island. Second, slow and steady lowering of the coverslip onto the TES fluorogel from one short end to the other short end will also decrease the likelihood of trapping air.

17. Due to its ubiquitous nature, the DAPI fluorescent signal will likely be more robust than the signal from West Nile virus envelope protein. Fluorescent West Nile virus envelope protein signal may need to be viewed at higher magnifications.

Acknowledgments

The authors thank Fengwei Bai and Farzana Nazneen at the University of Southern Mississippi for providing West Nile virus-infected tissue. This work was supported by the American Otological Society (DEV, KTY), the American Hearing Research Foundation (KTY) and the National Institute of General Medical Sciences of the National Institutes of Health under Award Number P20GM121334 (KTY). The content is solely the responsibility of the authors and does not necessarily represent the official views of the National Institutes of Health.

References

1. Yee KT, Neupane B, Bai F et al (2020) Zika virus infection causes widespread damage to the inner ear. Hear Res 395:108000. https://doi.org/10.1016/j.heares.2020.108000

2. Yee KT, Simon HH, Tessier-Lavigne M et al (1999) Extension of long leading processes and neuronal migration in the mammalian brain directed by the chemoattractant netrin-1. Neuron 24(3):607–622

3. Vetter DE, Adams JC, Mugnaini E (1991) Chemically distinct rat olivocochlear neurons. Synapse 7(1):21–43

4. Murthy V, Taranda J, Elgoyhen AB et al (2009) Activity of nAChRs containing alpha9 subunits modulates synapse stabilization via bidirectional signaling programs. Dev Neurobiol 69(14):931–949. https://doi.org/10.1002/dneu.20753

5. Murthy V, Maison SF, Taranda J et al (2009) SK2 channels are required for function and long-term survival of efferent synapses on mammalian outer hair cells. Mol Cell Neurosci 40(1):39–49. https://doi.org/10.1016/j.mcn.2008.08.011

6. Yee KT, Smetanka AM, Lund RD et al (1990) Differential expression of class I and class II major histocompatibility complex antigen in early postnatal rats. Brain Res 530(1):121–125. https://doi.org/10.1016/0006-8993(90)90667-z

7. Idikio HA (2009) Immunohistochemistry in diagnostic surgical pathology: contributions of protein life-cycle, use of evidence-based methods and data normalization on interpretation of immunohistochemical stains. Int J Clin Exp Pathol 3(2):169–176

8. von Behring E, Kitasato S (1890) Ueber das Zustandekommen der Diphtherie-Immunitata und der Tetanus-Immunitat bei Thieren. Deutsch Med Wochenschr 16:1113–1114

9. Heidelberger M, Kendall FE (1933) Studies on the precipitin reaction: precipitating haptens; species differences in antibodies. J Exp Med 57(3):373–379. https://doi.org/10.1084/jem.57.3.373

10. Marrack JR (1934) Derived antigens as a means of studying the relation of specific combination to chemical structure: (section of therapeutics and pharmacology). Proc R Soc Med 27(8):1063–1065

11. Coons A, Creech HJ, Jones RN (1941) Immunological properties of an antibody containing a fluorescent group. Proc Soc Exp Biol Med 47:200–202

12. Nakane PK, Pierce GB Jr (1966) Enzyme-labeled antibodies: preparation and application for the localization of antigens. J Histochem Cytochem 14(12):929–931. https://doi.org/10.1177/14.12.929

13. Childs GV (2014) History of immunohistochemistry. In: Pathology of human disease. Academic Press, Amsterdam, pp 3775–3796

14. Vetter DE, Cozzari C, Hartman BK et al (1993) Choline acetyltransferase in the rat cochlear nuclei: immunolocalization with a monoclonal antibody. In: Merchan MA, Juiz JM, Godfrey DA, Mugnaini E (eds) The mammalian cochlear nuclei: organization and function, NATO ASI series a: life sciences, vol 239. Plenum Press, New York, pp 279–290

15. Oertel WH, Schmechel DE, Mugnaini E et al (1981) Immunocytochemical localization of glutamate decarboxylase in rat cerebellum with a new antiserum. Neuroscience 6(12):2715–2735

16. Hsu SM, Raine L (1982) Versatility of biotin-labeled lectins and avidin-biotin-peroxidase complex for localization of carbohydrate in tissue sections. J Histochem Cytochem 30(2):157–161. https://doi.org/10.1177/30.2.7037937

17. Vetter DE, Liberman MC, Mann J et al (1999) Role of alpha9 nicotinic ACh receptor subunits in the development and function of cochlear efferent innervation. Neuron 23(1):93–103

18. Adams JC (1981) Heavy metal intensification of DAB-based HRP reaction product. J Histochem Cytochem 29(6):775–775. https://doi.org/10.1177/29.6.7252134

19. Adams JC (1992) Biotin amplification of biotin and horseradish peroxidase signals in histochemical stains. J Histochem Cytochem 40(10):1457–1463. https://doi.org/10.1177/40.10.1527370

20. Smithburn KC, Hughes TP, Burke AW et al (1940) A neurotropic virus isolated from the blood of a native of Uganda 1. Am J Trop Med Hyg s1-20(4):471–492. https://doi.org/10.4269/ajtmh.1940.s1-20.471

21. Hofmeister EK (2011) West Nile virus: North American experience. Integrative Zoology. https://doi.org/10.1111/j.1749-4877.2011.00251.x

22. Turell MJ, O'Guinn ML, Dohm DJ et al (2002) Vector competence of Culex tarsalis from Orange County, California, for West Nile virus. Vector Borne Zoonotic Dis https://doi.org/10.1089/15303660260613756

23. Bender K, Thompson FE (2003) West Nile Virus: A growing challenge. Am J Nurs 103(6):32–39. https://doi.org/10.1097/00000446-200306000-00018

24. Centers for Disease Control and Prevention (2003) Detection of West Nile virus in blood donations–United States, 2003. Morb Mortal Wkly Rep 52(32):769–772

25. Pealer LN, Marfin AA, Petersen LR et al (2003) Transmission of West Nile virus through blood transfusion in the United States in 2002. N Engl J Med 349(13):1236–1245. https://doi.org/10.1056/NEJMoa030969

26. Iwamoto M, Jernigan DB, Guasch A et al (2003) Transmission of West Nile Virus from an Organ Donor to Four Transplant Recipients. N Engl J Med 348(22):2196–2203. https://doi.org/10.1056/NEJMoa022987

27. Centers for Disease Control and Prevention (2002) Intrauterine West Nile virus infection–New York, 2002. Morb Mortal Wkly Rep 51(50):1135–1136

28. MMWR (2002) Possible West Nile virus transmission to an infant through breast-feeding—Michigan, 2002. Morb Mortal Wkly Rep 2002 515:877–878

29. Petersen LR, Brault AC, Nasci RS (2013) West Nile virus: Review of the literature. JAMA 310(3):308. https://doi.org/10.1001/jama.2013.8042

30. Bernard KA, Maffei JG, Jones SA et al (2001) West nile virus infection in birds and mosquitoes New York state 2000. Emerg Infect Dis 7(4):679–685. https://doi.org/10.3201/eid0704.010415

31. Oliveri RL, Jeter WC, Stark LM et al (2003) Surveillance results from the first West Nile virus transmission season in Florida 2001. Am J Trop Med Hyg 69(2):141–150. https://doi.org/10.4269/ajtmh.2003.69.141

32. Goddard LB, Roth AE, Reisen WK et al (2002) Vector competence of California mosquitoes for West Nile virus. Emerg Infect Dis 8(12):1385–1391. https://doi.org/10.3201/eid0812.020536

33. Blitvich BJ (2008) Transmission dynamics and changing epidemiology of West Nile virus. Anim Health Res Rev 9(1):71–86. https://doi.org/10.1017/S1466252307001430

34. Turell MJ, Dohm DJ, Sardelis MR et al (2005) An update on the potential of North American mosquitoes (Diptera: Culicidae) to transmit West Nile virus. J Med Entomol 42(1):57–62. https://doi.org/10.1093/jmedent/42.1.57

35. Turell MJ, O'Guinn ML, Dohm DJ et al (2001) Vector competence of North American mosquitoes (Diptera: Culicidae) for West Nile virus. J Med Entomol 38(2):130–134. https://doi.org/10.1603/0022-2585-38.2.130

36. Apperson CS, Hassan HK, Harrison BA et al (2004) Host feeding patterns of established and potential mosquito vectors of West Nile virus in the Eastern United States. Vector Borne Zoonotic Dis 4(1):71–82. https://doi.org/10.1089/153036604773083013

37. Hamer GL, Kitron UD, Brawn JD et al (2008) Culex pipiens (Diptera: Culicidae): A bridge vector of West Nile virus to humans. J Med Entomol 45(1):125–128. https://doi.org/10.1603/0022-2585(2008)45[125:CPDCAB]2.0.CO;2

38. Nasci RS, Savage HM, White DJ et al (2001) West Nile virus in overwintering Culex mosquitoes, New York City, 2000. Emerg Infect Dis 7(4):742–744. https://doi.org/10.3201/eid0704.010426

39. Komar N, Burns J, Dean C et al (2001) Serologic evidence for West Nile virus infection in birds in Staten island New York after an outbreak in 2000. Vector Borne Zoonotic Dis 1(3):191–196. https://doi.org/10.1089/153036601753552558

40. Komar N, Panella NA, Burns JE et al (2001) Serologic evidence for West Nile virus infection in birds in the New York City vicinity during an outbreak in 1999. Emerg Infect Dis 7(4):621–625. https://doi.org/10.3201/eid0704.010403

41. Komar N, Owen JC, Edwards E et al (2005) Avian hosts for West Nile virus in st. tammany Parish Louisiana 2002. Am J Trop Med Hyg 73(6):1031–1037. https://doi.org/10.4269/ajtmh.2005.73.1031

42. Kilpatrick AM, Kramer LD, Jones MJ et al (2006) West Nile virus epidemics in North America are driven by shifts in mosquito feeding behavior. PLoS Biol 4(4):e82. https://doi.org/10.1371/journal.pbio.0040082

43. Tempelis C, Reeves W, Bellamy R et al (1965) A three-year study of the feeding habits of Culex Tarsalis in Kern County, California. Am J Trop Med Hyg 14(1):170–177. https://doi.org/10.4269/ajtmh.1965.14.170

44. Francy DB, Tempelis CH, Hayes RO (1967) Variations in feeding patterns of seven culicine mosquitoes on vertebrate hosts in weld and larimer counties Colorado. Am J Trop Med Hyg 16(1):111–119. https://doi.org/10.4269/ajtmh.1967.16.111

45. Molaei G, Andreadis TG, Armstrong PM (2006) Host feeding patterns of Culex mosquitoes and West Nile virus transmission, Northeastern United States. Emerg Infect Dis 12(3):468–474. https://doi.org/10.3201/eid1203.051004

46. Campbell GL, Marfin AA, Lanciotti RS et al (2002) West Nile virus. Lancet Infect Dis 2(9):519–529.   S1473309902003687. https://doi.org/10.1016/S1473-3099(02)00368-7

47. Petersen LR, Roehrig RT (2001) West Nile virus: A reemerging global pathogen. Emerg Infect Dis 7(4):611–614. https://doi.org/10.3201/eid0704.010401

48. Zeller HG, Schuffenecker I (2004) West Nile virus: An overview of its spread in Europe and the mediterranean basin in contrast to its spread in the Americas. Eur J Clin Microbiol Infect Dis 23(3):147–156. https://doi.org/10.1007/s10096-003-1085-1

49. Chowers MY, Lang R, Nassar F et al (2001) Clinical characteristics of the West Nile fever outbreak Israel 2000. Emerg Infect Dis 7(4):675–678 https://doi.org/10.3201/eid0704.010414

50. Tsai TF, Popovici F, Cernescu C (1998) West Nile encephalitis epidemic in southeastern Romania. Lancet 352(9130):767–771. S0140673698035387. https://doi.org/10.1016/S0140-6736(98)03538-7

51. McIntosh BM, Jupp PG, dos Santos I et al (1976) Epidemics of West Nile and Sindbis viruses in South Africa with Culex (Culex) univittatus Thoebald as vector. S Afri J Sci 72:295–300

52. Parks JJ, Ganaway JR, Price WH (1958) Studies on immunologic overlap among certain arthropod-borne viruses. III. A laboratory analysis of three strains of West Nile virus which have been studied in human cancer patients. Am J Hyg 68(2):106–119

53. Nir Y, Goldwasser R, Lasowski Y et al (1968) Isolation of West Nile virus strains from mosquitoes in Israel. Am J Epidemiol 87(2):496–501. https://doi.org/10.1093/oxfordjournals.aje.a120839

54. Savage HM, Ceianu C, Nicolescu G et al (1999) Entomologic and avian investigations of an epidemic of West Nile fever in Romania in 1996 with serologic and molecular characterization of a virus isolate from mosquitoes. Am J Trop Med Hyg 61(4):600–611. https://doi.org/10.4269/ajtmh.1999.61.600

55. Lanciotti RS, Ebel GD, Deubel V et al (2002) Complete genome sequences and phylogenetic analysis of West Nile virus strains isolated from the United States Europe and the Middle East. Virology 298(1):96–105. S0042682202914492. https://doi.org/10.1006/viro.2002.1449

56. Danis K, Papa A, Theocharopoulos G et al (2011) Outbreak of West Nile virus infection in Greece 2010. Emerg Infect Dis 17(10):1868–1872. https://doi.org/10.3201/eid1710.110525

57. Papa A, Xanthopoulou K, Gewehr S et al (2011) Detection of West Nile virus lineage 2 in mosquitoes during a human outbreak in Greece. Clin Microbiol Infect 17 (8):1176–1180. S1198743X14629487. https://doi.org/10.1111/j.1469-0691.2010. 03438.x

58. Danis K, Papa A, Papanikolaou E et al (2011) Ongoing outbreak of West Nile virus infection in humans, Greece, July to August 2011. Euro Surveill 16(34):19951. https://doi.org/10. 2807/ese.16.34.19951-en

59. Anastasiadou A, Economopoulou A, Kakoulidis I et al (2011) Non-neuroinvasive West Nile virus infections during the outbreak in Greece. Clin Microbiol Infect 17(11):1681–1683. S1198743X14618978. https://doi.org/10. 1111/j.1469-0691.2011.03642.x

60. Fratkin JD, Leis AA, Stokic DS et al (2004) Spinal cord neuropathology in human West Nile virus infection. Arch Pathol Lab Med 128(5):533–537. https://doi.org/10.5858/ 2004-128-533-SCNIHW

61. Bai F, Thompson EA, Vig PJS et al (2019) Current understanding of West Nile virus clinical manifestations immune responses neuroinvasion and immunotherapeutic implications. Pathogens 8(4):193. https://doi.org/10. 3390/pathogens8040193

Chapter 8

Quantitative Analysis of B-Cell Subpopulations in Bone Marrow by Flow Cytometry

Tingting Geng and Penghua Wang

Abstract

Flow cytometry is a technology that rapidly detects and measures physical and chemical characteristics of single cells or particles. It is a powerful tool for many areas of research, in particular, immunology, which allows for simultaneous analysis of different immune cell populations in a tissue. Here we describe the procedures to quantify and/or purify various B fractions in mouse bone marrows by flow cytometry using their signature surface markers. This method is useful to study B-cell development during steady-state or emergency hematopoiesis such as viral infections.

Key words B cell, Flow cytometry, Lymphopoiesis, Subpopulation

1 Introduction

The mammalian immune system consists of many distinct immune cell populations that coordinate responses to infection, maintaining tissue and immune homeostasis. Blood immune cells are constitutively differentiated from bone marrow stem cells through progenitor cells [1], a process known as hematopoiesis. In response to microbial infections, hematopoiesis is rapidly mobilized, generally beginning with myeloid lineages (neutrophils, monocytes, etc.) and then lymphoid cells. The hematopoietic system is a hierarchically organized, somatic stem cell-maintained organ system, with long-lived and self-renewing pluripotent hematopoietic stem cells (LT-HSCs) at its apex [1]. LT-HSCs differentiate into short-term multipotent progenitors and lineage-committed hematopoietic progenitors, which in turn differentiate into mature blood cell lineages [2]. B cells, along with T cells, form the core of the adaptive immune arm that is essential for microbial clearance. However, B cells are active participants in pathogenesis of numerous autoimmune diseases, including systemic lupus erythematosus and multiple sclerosis [3]. B-cell-depleting therapies are thus being

Fengwei Bai (ed.), *West Nile Virus: Methods and Protocols*,
Methods in Molecular Biology, vol. 2585, https://doi.org/10.1007/978-1-0716-2760-0_8,

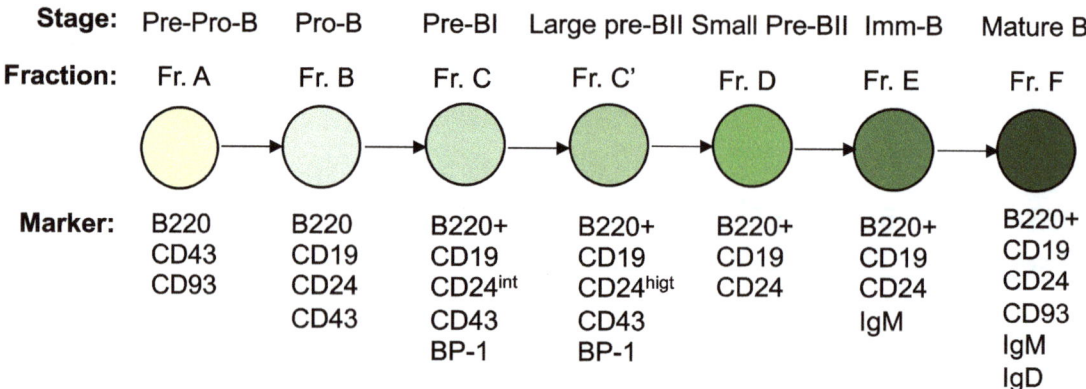

Fig. 1 A schematic diagram of defined stages of B-cell development

actively pursued with promising clinical success. Inside the bone marrow, B lymphopoiesis starts from common lymphoid progenitors, then B lineage committed pre-progenitor (pre-pro) B, pro-B, precursor B I (pre-BI), large pre-BII, small pre-BII, immature B (imm-B), and mature (recirculating) B [4].

Flow cytometry is a technology that rapidly detects and measures physical and chemical characteristics of single cells or particles. Flow cytometers utilize lasers to produce both scattered and fluorescent light signals that are sensed and quantitated in photodiodes or photomultiplier tubes. Flow cytometry is a powerful tool for many areas of research, in particular, immunology, which allows for simultaneous analysis of different immune cell populations in any tissue. Cell populations are stained by fluorophore-conjugated antibodies against their unique surface markers, and then quantified and/or purified based on their fluorescent or light scattering characteristics by a flow cytometer [5]. Here, we describe the procedures to quantify and/or purify various B fractions in mouse bone marrows by flow cytometry using their signature surface markers (Fig. 1). This method is useful to study B-cell development during steady-state or viral infections.

2 Materials

1. 1.2 mL bullet tube.

2. 25 G needle.

3. Autoclaved 0.5 mL and 1.5 mL Eppendorf microfuge tubes and 15 mL and 50 mL Falcon conical tubes.

4. Phosphate buffered saline (PBS): 1× PBS pH 7.4 Gibco, Catalogue No. 10010-023.

5. PBS/ethylenediaminetetraacetic acid (EDTA): 1× PBS pH 7.4 with 1 mM EDTA.

Table 1
Fluorophore-conjugated antibody mix for B-lineage surface stain

Antibody	Fluorophore	Supplier	Catalog	Volume (μL)
CD3	FITC	BioLegend	100204	1
CD11b	FITC	BioLegend	101206	1
LY6G/6C	FITC	BioLegend	108406	1
TER-119	FITC	BioLegend	116206	1
F4/80	FITC	BioLegend	123108	1
CD19	PE/Dazzle™ 594	BioLegend	115554	1
CD45R	Brilliant Violet 650	BioLegend	103241	1
IgM	Brilliant Violet 421	BioLegend	406532	1
CD93	PE-Cy7	BioLegend	136506	1
IgD	APC-Cy7	BioLegend	405716	1
CD43	APC	BioLegend	121214	1
CD24	PerCP-Cy5.5	BioLegend	101824	1
BP-1	PE	BioLegend	108308	1
TruStain FcX PLUS		BioLegend	156604	1
Brilliant Stain Buffer Plus		BD Bioscience	566385	10
FACs buffer				76
Total volume				100

6. Rodent red blood cell (RBC) lysis buffer: Alfa Aesar, Catalogue No. J62150.

7. Fluorescence-activated cell sorting (FACS) buffer: $1\times$ PBS pH 7.4 with 1 mM of EDTA and 2% fetal bovine serum (FBS).

8. Zombie UV™ solution: Pre-warm the Zombie UV™ dye kit (BioLegend, Catalogue No. 877-246-5343) to room temperature; add 100 μl of dimethylsulfoxide (DMSO) to one vial of Zombie UV™ dye, and dissolve it completely. Dilute Zombie UV™ dye at 1:100 in PBS/EDTA before use.

9. B-lineage stain antibody mix: *See* Table 1 for the cocktail ingredients. The antibody mix contains TruStain FcX Plus anti-CD16/32 (BioLegend, Catalogue No. 101319) to reduce nonspecific binding of fluorescent antibodies and Brilliant Stain Buffer Plus (BD Biosciences, Catalogue No. 566385) to increase the performance of Brilliant Violet.

10. UltraComp eBeads: Invitrogen, Catalogue No. 01-222-42.

11. Mice: >6 weeks old C57BL/6.

12. Carbon dioxide (CO_2) euthanasia chamber.

13. Swing bucket refrigerated centrifuge, fix-angled refrigerated microfuge.

14. Sterilized surgical scissors and scalpel.

15. BD LSR II flow cytometer.

3 Methods

3.1 Preparation of Single-Cell Suspension

1. Punch two holes on the bottom of a 0.5 mL microfuge tube with a 25 G needle. Insert this 0.5 mL microfuge tube into a 1.5 mL microfuge tube. This 0.5−/1.5 mL tube set is for isolation of the bone marrow.

2. Euthanize a mouse in a CO_2 chamber plus cervical dislocation per animal protocol; disinfect the mouse with 70% ethanol. Peel the skin from the top of each hind leg and down over the foot. Cut off the foot along with the skin and discard. Cut off the hind leg at the hip joint with scissors, being careful to leave the femur intact. Place both the femurs and tibias in a 60 mm Petri dish containing 5 mL of ice-cold PBS.

3. Remove excess muscle and skin from the leg by holding one end of the bone with a tweezers and using scissors to push muscle downward away from the tweezers. Carefully sever leg bones proximal to each joint.

4. Cut off one end of each of the femur and tibia. Place them inside the 0.5/1.5 microcentrifuge tube set with the cut side down. Immediately centrifuge at $6000 \times g$ for 15 s in a microfuge at 4 °C. All the bone marrow should be centrifuged to the bottom of the 1.5 mL tube. Keep the bone marrow on ice while processing other specimens.

5. Transfer and resuspend the bone marrow in 3 mL RBC lysis buffer (kept at room temperature) in a 15 mL conical tube. Incubate for 3 min at room temperature. Stop the lysis reaction by adding 10 mL of ice-cold $1\times$ PBS/EDTA. Centrifuge at $500 \times g$ for 5 min at 4 °C and decant the supernatant. Resuspend cell pellets in 4 mL of ice-cold PBS/EDTA and filter through a 70 μm strainer to remove aggregated cell clumps and debris.

6. Count cell (*see* **Note 1**). Keep the sample tube on ice.

3.2 Cell Staining

1. Transfer ~1×10^6 cells to a 1.5 mL microfuge tube. Centrifuge at $400 \times g$ for 5 min at 4 °C and remove the supernatant.

2. Add 100 µL Zombie UV™ solution. Gently mix and incubate on ice, in the dark, for 20 min. Add 1 mL of ice-cold FACS buffer. Centrifuge at $400 \times g$ for 5 min at 4 °C and remove the supernatant.

3. Add 100 µL of B-lineage stain antibody mix with Brilliant Stain Buffer (*see* **Note 2**). Gently mix and incubate on ice, in the dark, for 30 min.

4. Add 1 mL ice-cold FACS buffer. Centrifuge at $400 \times g$ for 5 min in 4 °C and remove the supernatant.

5. Repeat **step 4** once.

6. Resuspend cells in 300 pi1;µL ice-cold FACS buffer. Filter the cells through a 40 µm mesh to a 1.2 mL bullet tube. Keep the sample on ice covered with aluminum foil until flow cytometry analysis.

3.3 Preparation of Compensation Beads

1. Resuspend ~1×10^5 unstained cell in 300 µL ice-cold FACS buffer in a 1.5 mL microfuge tube. Filter the cells through a 40 µm mesh to a 1.2 mL bullet tube, and mark it "Unstained".

2. Vortex UltraComp eBeads for 10 s.

3. Add the compensation beads into ice-cold FACS buffer (300 µL FACS buffer for each color. One drop for every 4–5 colors of antibody).

4. Mix well and aliquot 300 µL of beads to each bullet tube.

5. Add 0.5 µL of each fluorophore-conjugated antibody to one tube. Vortex and mix. Mark the color name on the tube.

6. Keep all the compensation tubes on ice covered with aluminum foil till flow cytometry analysis.

3.4 Flow Cytometry and Data Analysis with FlowJo

1. Run samples on a BD LSR II Flow Cytometer (*see* **Note 3**). Run "unstained" control and compensation beads control for each fluorophore-conjugated antibody from Subheading 3.3. After compensation calculation, appropriately set the gates and record all the samples on the flow cytometer.

2. Analyze data using the FlowJo software 10.8.1. Firstly, gate single and live cells. Secondly, dump non-B cells positive for FITC-CD3, -TER119, -LY6G/6C, -CD11b, and F4/80 (Dump). Thirdly, gate the remaining cells sequentially on BV650 anti-B220, APC anti-CD43, PerCP-Cy5 anti-CD24, PE anti-BP1/CD249/Ly51, BV421 anti-IgM, APC-Cy7 anti-IgD, PE-Cy7 anti-CD93, and PE-Dazzle 594 anti-CD19. See Fig. 2 for an example of the gating strategy.

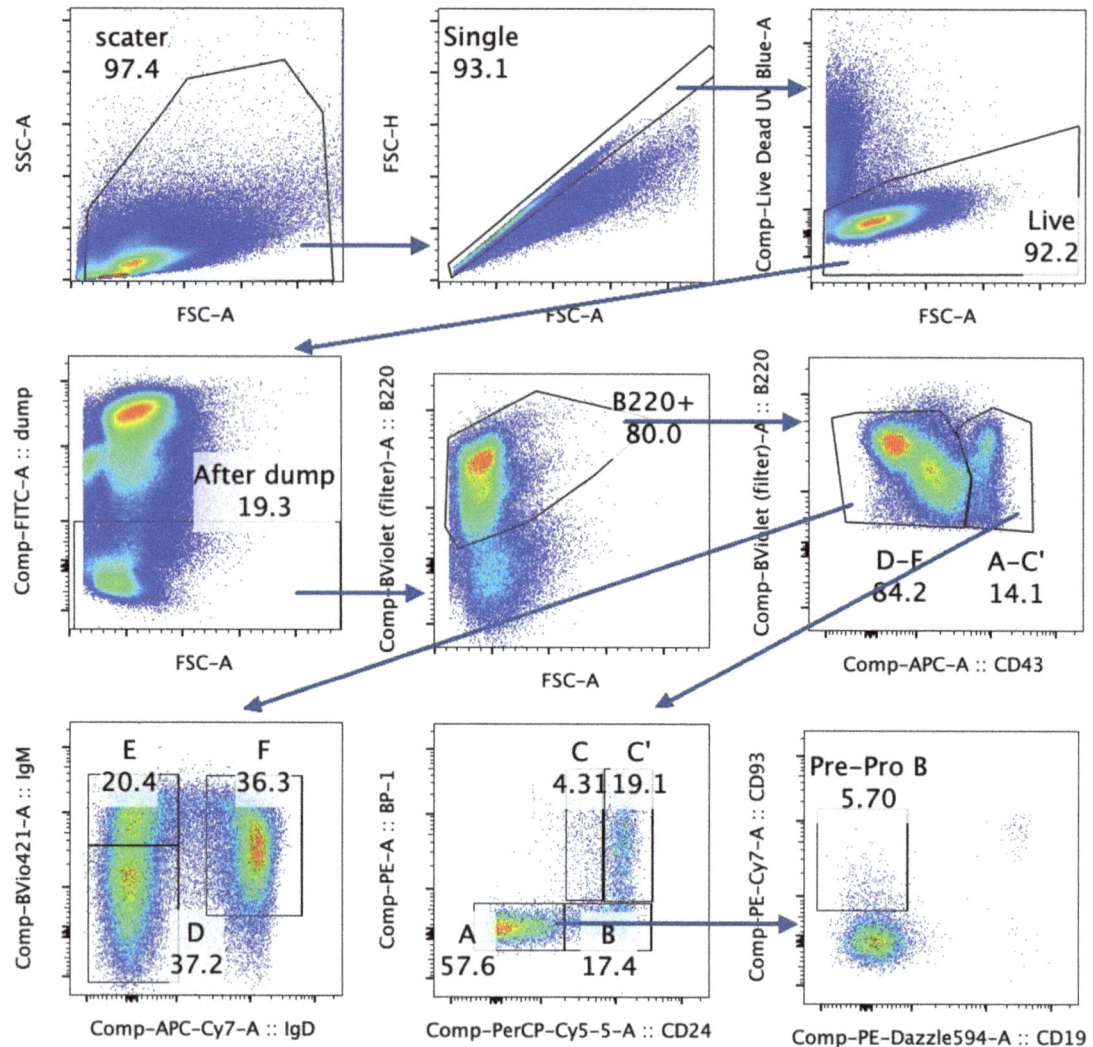

Fig. 2 The sequential gating strategy. Total live single cells were gated on a non-B cocktail (Dump). The Dump negative population was sorted into B220$^+$cells, which were then gated into CD43^{hi} (Fraction A-C') and CD43$^{Lo/-}$ (Fraction D-F) cells. Fraction A-C' was sorted into BP-1$^+$CD24$^{Low\ (Lo)}$ precursor (pre)-BI (Fraction C), BP-1$^+$CD24$^{High\ (hi)}$ large (L) pre-BII (Fraction C'), BP-1$^-$CD24$^-$ Fraction A, and BP-1$^-$CD24$^+$ progenitor (pro)-B (Fraction B). Fraction A was further gated into CD19$^-$ CD93$^+$ pre-progenitor (pre-pro-B). Fraction D-F was sorted into IgD$^-$IgM$^-$ small pre-BII (Fraction D), IgD$^-$ IgM$^+$ immature (Imm) B (Fraction E), and IgD$^+$ IgM$^+$ mature B (Fraction F). The arrows indicate the sequential sorting order

4   Notes

1. We use a Bio-Rad TC20-automated cell counter. Alternatively, a conventional glass hemocytometer together with a trypan blue stain will work well.

2. BD Horizon Brilliant Stain Buffer Plus should be used in multicolor immunofluorescent staining when using two or

more BD Horizon Brilliant reagents. The 10 μL amount of Brilliant Stain Buffer Plus per tube does not depend on the final staining volume or amount of cells used per tube or number of BD fluorescent antibodies used in staining.

3. The flow cytometer used in this protocol was the LSR-II flow cytometer, equipped with 4 lasers (355 nm UV, 405 nm violet, 488 nm blue, 561 nm yellow-green) and 14 channels supported.

Acknowledgments

This work was supported by a US National Institutes of Health research award NIAID R01AI132526 to P.W.

Conflict of Interest The authors declare no financial or nonfinancial interest.

References

1. Boettcher S, Manz MG (2017) Regulation of inflammation- and infection-driven hematopoiesis. Trends Immunol 38:345–357

2. Rieger MA, Schroeder T (2012) Hematopoiesis. Cold Spring Harb Perspect Biol 4:a008250

3. Lee DSW, Rojas OL, Gommerman JL (2021) B cell depletion therapies in autoimmune disease: advances and mechanistic insights. Nat Rev Drug Discov 20:179–199

4. Martensson IL, Almqvist N, Grimsholm O et al (2010) The pre-B cell receptor checkpoint. FEBS Lett 584:2572–2579

5. Mckinnon KM (2018) Flow cytometry: an overview. Curr Protoc Immunol 120:5.1.1–5.1.11

Chapter 9

Isolation of Exosomes or Extracellular Vesicles from West Nile Virus-Infected N2a Cells, Primary Cortical Neurons, and Brain Tissues

Hameeda Sultana and Girish Neelakanta

Abstract

Several flaviviruses compromise the blood-brain barrier integrity, infect the central nervous system, and elicit neuroinvasion to successfully cause neuropathogenesis in the vertebrate host. Therefore, understanding the pathway(s) and mechanism(s) to block the transmission and/or dissemination of flaviviruses and perhaps other neuroinvasive viruses is considered as an important area of research. Moreover, studies that address mechanism(s) of neuroinvasion by flaviviruses are limited. In this chapter, we discuss detailed methods to isolate exosomes or extracellular vesicles (EVs) from mouse and human N2a cells, primary cultures of murine cortical neurons, and mouse brain tissue. Two different methods including differential ultracentrifugation and density gradient exosome (DG-Exo) isolation are described for the preparation of exosomes/EVs from N2a cells and cortical neurons. In addition, we discuss the detailed DG-Exo method for the isolation of exosomes from murine brain tissue. Studies on neuronal exosomes will perhaps enhance our understanding of the mechanism of neuroinvasion by these deadly viruses.

Key words West Nile virus, N2a cells, Cortical neurons, Mice brain, Transmission, Exosomes, Extracellular vesicles

1 Introduction

Understanding the process and mechanism(s) of neuroinvasion by flaviviruses has to be taken into serious consideration. Several of the mosquito- and tick-borne flaviviruses such as West Nile virus (WNV), Japanese encephalitis virus (JEV), St. Louis encephalitis virus (SLEV), Zika virus (ZIKV), tick-borne encephalitis virus (TBEV), and tick-borne Powassan virus (POWV) that infect humans and other vertebrate hosts, including our pets such as horses and dogs, are neuroinvasive viruses [1–13]. These neurotropic flaviviruses are emerging and re-emerging pathogens that are responsible for the increasingly severe encephalitis outbreaks in humans and animals [3, 14–16]. The neuroinvasion of these

Fengwei Bai (ed.), *West Nile Virus: Methods and Protocols*,
Methods in Molecular Biology, vol. 2585, https://doi.org/10.1007/978-1-0716-2760-0_9,

flaviviruses causes neuropathogenesis, neuronal loss, neuroinflammation, and encephalitis that ultimately lead to the death of the host [1, 4, 7, 9, 10, 16–22]. How these neuroinvasive viruses trick and gain entry into the brain is still poorly understood and is mostly based on hypothesis-driven discussions. So far, no studies clearly show the mechanism(s) of how these flaviviruses are transmitted from one tissue to the other in the periphery and then disseminated into the central nervous system (CNS) of humans and animals. Several studies have focused on the blood-brain barrier (BBB) permeability and compromised systems that allow the infection of the brain [16, 22–29]. These studies reveal the transneural or Trojan horse mechanisms for the entry of flaviviruses that revolves around the BBB disruption followed by retrograde axonal transport [16, 22–29]. In our recent studies, we have provided a proof of principle concept that neuroinvasion of flaviviruses is perhaps mediated via extreme production, release, and dissemination of exosomes or extracellular vesicles (EVs, small-membrane-bound vesicles) in the peripheral system that may breach the BBB integrity leading to the infection of the CNS [30, 31]. Therefore, studies on exosomes and their isolation methods are highly important to understand the neurotransmission of flaviviruses into the brain. Our published studies with extensive analysis have shown that exosomes/EVs can be isolated from the cell culture supernatants and also from the peripheral tissues [30–34]. Studies with these exosomes/EVs provided evidences to understand the pathogen-driven mechanisms and neuroinvasion into the brain [30–34]. We assume that exosomes/EVs produced and released by the peripheral system perhaps allow dissemination into the cerebrospinal fluid (CSF) for an en-route to the brain. Our recently published studies showed that exosomes/EVs isolated from infected mouse neuronal cell lines (N2a) or primary in vitro cultures of cortical neurons had abundant loads of both tick-borne Langat virus (LGTV) or mosquito-borne WNV or ZIKV RNA, and proteins [30, 31]. In addition, we have also reported the isolation of exosomes/EVs from the brain tissue [32]. In the murine model, it has been shown that at an early stage, WNV replicates in the blood (at days 1–2 postinfection, p.i.), thereby increasing the viremia that allows the viral dissemination and replication in the peripheral organs such as the spleen and liver (at days 2–4 p.i.), and then cross the BBB (at days 5–6 p.i.) to infect the CNS (at days 6–8 p.i.), thereby leading to the death of mice within 2 weeks (days 7–14 p.i.) [1, 14, 16, 21, 22, 28, 35]. We believe that flaviviral entry into the vertebrate peripheral system happens via secretion of arthropod infectious exosomes (presented in saliva from ticks/mosquitoes) that leads to the fusion of these vesicles on recipient cells via receptor-mediated endocytosis. The secretion of infectious exosomes from variety of cells in the periphery could then lead to the compromised BBB and entry of virus into the CNS (Fig. 1).

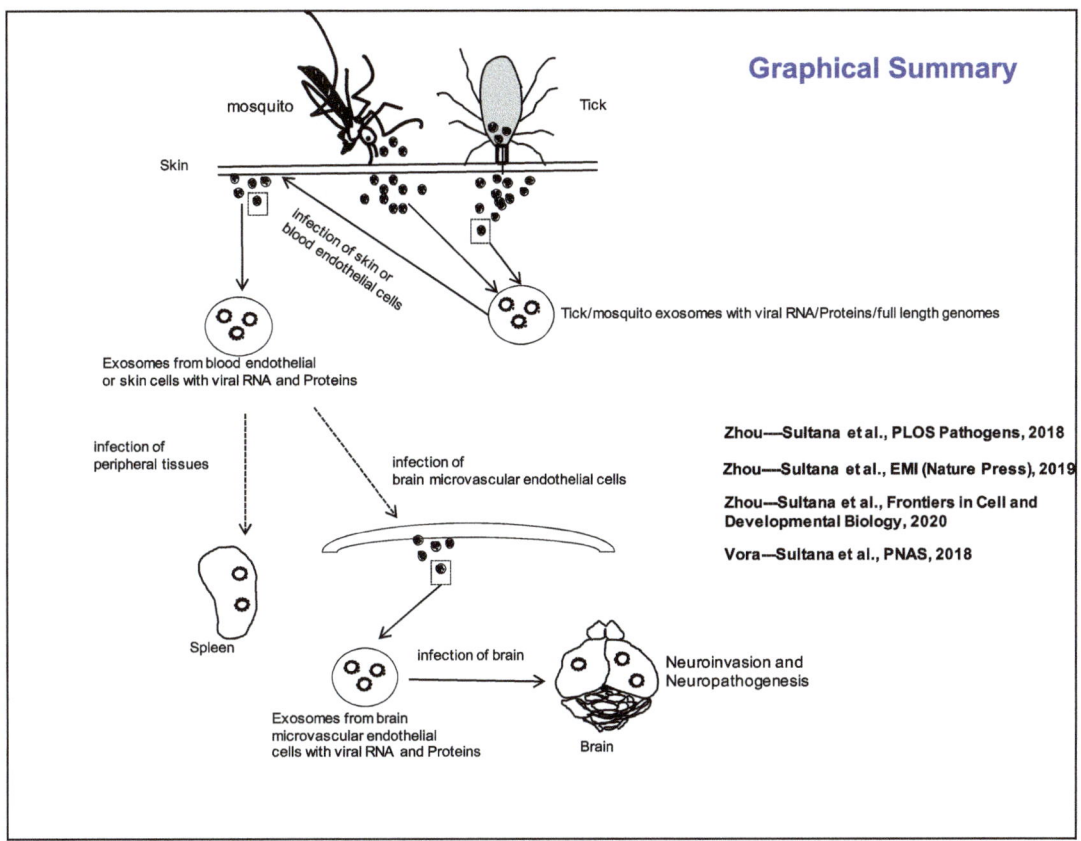

Fig. 1 Schematic representation showing tick-/mosquito-borne flaviviruses transmission to the vertebrate host via arthropod exosomes. The bite from mosquitoes or ticks would deposit infectious exosomes that perhaps fuse with the naïve recipient skin keratinocytes. This could trigger the production and release of exosomes/EVs to disseminate into the endothelial barrier and further lead this infection into the peripheral system including the tissues such as the spleen. Higher viremia/viral loads in peripheral tissues eventually allows breaching and entry/replication of the flaviviruses in to the brain microvascular endothelial cells that line the blood-brain barrier (BBB). Early replication of flaviviruses in the brain microvascular endothelial cells could result in fast production and release of infectious exosomes/EVs that would allow these viruses to infect neurons. Being highly communicable cells, neurons would produce excessive numbers of infectious exosomes/EVs and quickly turn around the infection via dissemination to naïve neighboring neurons, thereby causing severe neuropathogenesis and neuronal loss

This mode of exosomal-mediated viral entry not only provides an emerging concept to understand how flaviviruses transmit from vector to the vertebrate host but also leads us in understanding their dissemination strategies within the host. This chapter highlights the methods for the isolation of exosomes/infectious exosomes or EVs from neuronal cultures and the murine brain tissue.

2 Materials

Recently, the risk assessment for work with WNV was revised to indicate requirement of a biosafety level II laboratory containment with proper work practices. Individuals must still be thoroughly trained with required practices and orientations (please refer to the sixth edition of BMBL, https://www.cdc.gov/labs/BMBL.html). An authorization at the BSL-2 level is required for the institutional/laboratory authorities and with approval of the required protocols (*see* **Note 1**).

2.1 Isolation of Exosomes/EVs from Neuronal Cultures

2.1.1 Isolation of Exosomes/EVs from In Vitro Cultures of Mouse or Human N2a Cell Lines by Differential Ultracentrifugation Method

1. ATCC cell lines (mouse neuro-2a; catalog number CCL-131; human N2a cells; SH-SY5Y; catalog number CRL-2266).

2. Eagle's Minimum Essential Medium (EMEM, ATCC catalog number 30–2003).

3. Fetal bovine serum (FBS).

4. Trypsin-EDTA solution (1×).

5. Exosome-free FBS (Systems BioSciences Inc.).

6. Phosphate buffered saline (PBS, 1×).

7. Tissue culture flask, sterile and treated (*see* **Note 2**).

8. Penicillin (10 units/mL) and streptomycin (1 μg/mL) mixture.

9. Gloves.

10. Serological pipettes, sterile.

11. Pipettes and pipette tips, sterile.

12. Laboratory virus stocks with known titers (*see* **Note 3**).

13. Biosafety cabinet.

14. CO_2 incubator, set at 37 °C.

15. Inverted microscope.

16. Filtering devices (0.1 μm), VACUCAP filter for conical tubes.

17. Corning Spin-X UF concentrators or centrifugal filter device with a 5 K nominal molecular weight limit (NMWL), Spin-X UF 20 for samples up to 20 mL, catalog Number CLS431487, Millipore Sigma (Recommended) (*see* **Note 4**).

18. Benchtop centrifugation units, refrigerated.

19. Centrifuge that includes swinging-bucket rotor, buckets with lids, and adapters (for 15 and 50 mL tubes).

20. Centrifugation tubes (15 and 50 mL), with flat caps, conical bottom, polypropylene, sterile.

21. Complete media (EMEM media with 10% FBS and 5% penicillin (10 units/mL) and streptomycin (1 μg/mL) mixture).

22. Optima XPN or XE ultracentrifuge floor unit (Beckman Coulter, recommended) including appropriate rotors to spin volumes from 5 to 50 mL (*see* **Note 5**).

23. Ultracentrifugation tubes (Beckman Coulter, recommended) (*see* **Note 6**).

24. Standard-duty plastic cart (*see* **Note 7**).

25. Modified RIPA buffer (*see* **Note 8**).

2.1.2 Isolation of Exosomes/EVs from In Vitro Primary Cultures of Murine Cortical Neurons by Differential Ultracentrifugation Method

1. C57BL/6 wild-type gestation female mice of age (E12–13) (*see* **Notes 9** and **10**).

2. Culture plates of sizes 60 and 100 mm, sterile, nontreated/treated.

3. Poly-L-Lysine or Poly-L-Ornithine solution (0.01%).

4. Dissection buffer (25 mM of HEPES with 0.5% glucose).

5. Multiwell dishes (12- or 6-well culture plates).

6. Tissue culture flask, sterile and treated.

7. Neurobasal medium, liquid $1\times$ (Invitrogen, Thermo Fisher Scientific, recommended).

8. Heat-inactivated FBS.

9. B-27 supplement (2%) (Invitrogen, Thermo Fisher Scientific, recommended).

10. Glutamax/glutamine (2 mM).

11. Neurobasal complete media (neurobasal medium with 5% FBS, 2 mM Glutamax/L-glutamine, 2% B-27 supplement, and 5% penicillin (10 units/mL) and streptomycin (1 μg/mL) mixture).

12. Benchtop centrifugation units, refrigerated.

13. Centrifuge that includes swinging-bucket rotor, buckets with lids, and adapters (for 15 and 50 mL tubes).

14. Centrifugation tubes (15 and 50 mL), with flat caps, conical bottom, polypropylene, sterile.

15. Optima XPN or XE ultracentrifuge floor units (Beckman Coulter, recommended) including appropriate rotors to spin volumes from 5 to 50 mL (*see* **Note 5**).

16. Ultracentrifugation tubes (*see* **Note 6**).

17. Standard-duty plastic cart (*see* **Note 7**).

18. Modified RIPA buffer (*see* **Note 8**).

19. DR Instruments Deluxe Dissecting Kit (with microknife, Fisher Scientific, catalog number S06840, recommended).

20. HEPES solution (1 M, pH 7.0–7.6, sterile/filtered).

21. Glucose solution (0.5% and 36% glucose dissolved in ultrapure water).

22. NeuroCult™ Chemical Dissociation Kit (mouse; STEMCELL Technologies, catalog number 05707, recommended).

23. Certified fume hood.

24. Materials 8–17 from Subheading 2.1.1 are also needed for isolation of exosomes from cortical neurons.

2.1.3 Isolation of Exosomes/EVs from In Vitro Cultures of Mouse or Human N2a Cell Line or Primary Cultures of Murine Cortical Neurons by Density Gradient (DG-Exo) Method

In addition to all the materials described above in Subheadings 2.1.1 or 2.1.2, respectively, the following are the two additional items needed for DG-Exo method.

1. OptiPrep™ density gradient medium, 60% (w/v) of aqueous iodixanol (Millipore Sigma, cat number D1556-250ML, recommended).

2. 0.25 M Sucrose/10 mM Tris–HCl, pH 7.5.

2.1.4 Isolation of Exosomes/EVs from Mouse Brains by Density Gradient (DG-Exo) Method

1. C57BL/6 wild-type mice.

2. DR Instruments Deluxe Dissecting Kit (with microknife).

3. Culture plates of sizes 60 and 100 mm, sterile, nontreated/treated.

4. Mice brain slices of ~20 mg (fresh or frozen).

5. Weighing scale (analytical balance is preferred).

6. Razor blades (*see* **Note 11**).

7. U/mL of collagenase type III (STEMCELL Technologies, catalog number 07422_C, recommended).

8. Hibernate™-E Medium (Thermo Fisher Scientific, catalog number A1247601, recommended).

9. Shaking water bath.

10. Protease and Phosphatase inhibitors cocktail (ready-to-use).

11. Filtering devices (0.1 μm), VACUCAP filter for conical tubes.

12. Corning Spin-X UF concentrators or centrifugal filter device with a 5 K nominal molecular weight limit (NMWL), Spin-X UF 20 for samples up to 20 mL (*see* **Note 4**).

13. Centrifuge that includes Swinging-bucket Rotor, Buckets with lids, and adapters (for 15 and 50 mL tubes).

14. Centrifugation tubes (15 and 50 mL), with flat caps, conical bottom, polypropylene, sterile.

15. Optima XPN or XE ultracentrifuge floor units including appropriate rotors to spin volumes from 5 to 50 mL (*see* **Note 5**).

16. Ultracentrifugation tubes (*see* **Note 6**).

17. Standard-duty plastic cart (*see* **Note 7**).

18. OptiPrep™ density gradient medium, 60% (w/v) of aqueous iodixanol, cat number D1556-250ML, 250 ML, Millipore Sigma (recommended).

19. 0.25 M sucrose/10 mM Tris–HCl, pH 7.5.

20. Phosphate buffered saline (PBS, 1×).

3   Methods

3.1   Isolation of Exosomes/EVs from In Vitro Cultures of Mouse or Human N2a Cell Lines by Differential Ultracentrifugation Method

1. N2a cells are plated in complete media and to a desired density (*see* **Notes 12** and **13**).

2. Allow overnight incubation of cells for adherence (by incubations at 37 °C and 5% CO_2). If bovine exosomes have to be depleted, replace the complete media with 5–10% of bovine exosome-free FBS (exo-free FBS). Incubate cells in exo-free FBS for 4 h, and infect with WNV (e.g., CT2741 wild-type strain; with desired multiplication of infection: MOI) (*see* **Note 14**).

3. After desired days of postinfection (e.g., 1–5 days), collect the cell culture supernatant in 1.5–2 mL or 15–50 mL tubes based on volumes (*see* **Note 4**).

4. Cells can be visualized (using the inverted microscope with phase contrast) for estimating density and can be collected for RNA/protein extractions.

5. For isolation of exosome/EVs by Differential ultracentrifugation method, plate neuronal cells at desired density. Briefly, cell culture supernatants are spun at $300 \times g$ for 10 min, cell pellet is discarded, and the clear supernatant is spun again at $2000 \times g$ for 10 min. The pellet containing dead cells is discarded, and the supernatant is spun again at $10,000 \times g$ for 30 min to remove cell debris (*see* **Note 15**).

6. The supernatants collected from this above step are transferred to respective ultracentrifugation tubes, and the volumes are adjusted accordingly (with similar media, in which the supernatants are collected in) to balance the tubes (*see* **Note 16**).

7. The chilled ultracentrifugation rotor is directly brought to the inside of the biosafety cabinet, and the ultracentrifugation tubes are arranged in balancing order. The rotor cap is secured tight, and the rotor is loaded on a standard/heavy-duty cart and transported to the ultracentrifugation unit to spin at $100,000 \times g$ for 70 min (*see* **Notes 7** and **17**).

8. The supernatants are collected in 15–50-mL regular tubes and kept aside (*do not* discard these cell culture supernatants, as they can used as exosome depleted supernatants). The pellet

containing exosomes and any contaminants (in ultracentrifugation tubes) is washed/resuspended with ice-cold 1× PBS and spun again at 100,000×*g* for 70 min (*see* **Note 18**).

9. The resulting pellet of exosomes is considered as exosomal fraction. The pellet of exosomes is reconstituted in 1× PBS. Purified exosome preparations can be used immediately or stored at −80 °C and used later for further analysis (*see* **Note 19**).

3.2 Isolation of Exosomes/EVs from In Vitro Primary Cultures of Murine Cortical Neurons by Differential Ultracentrifugation Method

1. C57BL/6 wild-type mice with gestation period of E11–13 are obtained (from either in-bred laboratory colony or from a commercial vendor), and mice are allowed to reacclimatize for 2–3 days, based on the approvals from IACUC (*see* **Notes 9 and 10**). For isolation of primary cultures of cortical neurons, embryonic days 15/16 (E15/E16) is needed, and therefore the mice are acclimatized until this time.

2. Animals are transported into a research laboratory on the day of dissection (E15/16), for easy collection of embryonic brains and handling of tissues and culturing of neurons in a sterile cell culture facility. Mouse cages are transported in secondary large and secured containers with little to no obstructions during the transportation (*see* **Note 20**).

3. On day E15/16, pregnant female mice are euthanized according to the IACUC approved method. If euthanasia is performed in a research laboratory, a certified fume hood is required (*see* **Note 21**).

4. After euthanasia, the mouse fur is sprayed with ethanol (for sterilization), and the uterine horn is cut (about 8–10 mm) with small scissors provided in the dissection kit.

5. Embryos are collected by teasing the placental sac and immediately soaked in the cold solution of dissection buffer. Each embryonic brain is collected (one-by-one) in the fresh dissection buffer (use 60 mm plates).

6. The skin and cartilage are peeled and the brain is removed from the midbrain forward. Using a microknife, the telencephalic hemispheres are sliced off, and the olfactory bulbs are removed. The pia mater is stripped off and the dorsal telencephalon is dissected from the basal ganglion. The tissue is minced off and incubated in the nonenzymatic NeuroCult chemical disassociation solution (mouse).

7. Neurons are disassociated as per the NeuroCult chemical disassociation method (*see* **Note 22**).

8. The disassociated cells are collected in neurobasal complete media.

9. Neurons are counted either manually using a hemocytometer cell counter or by using the automated cell counter instrument.

10. Plate neurons at desired density and in complete medium (plates are coated with either Poly-L-Lysine (PLL) or Poly-L-Ornithine (PLO) 0.01% solution for 4 h). Plating neurons on PLL or PLO is very much recommended to allow better adherence of these cells (*see* **Note 23**).

11. After 1-day of post-plating, half of the complete media is replaced with serum depleted media (with 2.5% FBS) to allow neuronal growth and diminution of glial cells.

12. After replacement with depleted FBS media, neurons are infected with WNV or other desired flaviviruses (e.g., at desired MOI of 2).

13. Exosomes/EVs are isolated on day 3 postinfection or on a desired day after infection. After desired days of postinfection, collect the cell culture supernatant in 1.5–2 mL or 15–50 mL tubes based on volumes. If desired, the supernatants are filtered using the filtering devices (of 0.1 μm). Also, if higher volumes of supernatants are collected, then samples can be concentrated using the Corning Spin-X UF concentrators or centrifugal filter device with a 5–10 K nominal molecular weight limit (NMWL) (*see* **Note 4**).

14. Cells can be visualized (using the inverted microscope with phase contrast) for estimation of density and collected for RNA/protein extractions.

15. For isolation of exosome/EVs by Differential ultracentrifugation method, plate neuronal cells at desired density. Briefly, cell culture supernatants are spun at $300 \times g$ for 10 min, cell pellet was discarded, and the clear supernatant are spun again at $2000 \times g$ for 10 min. The pellet containing dead cells is discarded, and the supernatant is spun again at $10,000 \times g$ for 30 min to remove cell debris (*see* **Note 15**).

16. The supernatants collected from this above step are transferred to respective ultracentrifugation tubes (based on the availability of rotors in respective facilities), and the volumes are adjusted accordingly (with media in which the supernatants are collected) to balance the tubes (*see* **Note 16**).

17. The chilled ultracentrifugation rotor is directly brought to the inside of the biosafety cabinet, and the ultracentrifugation tubes are arranged in balancing order. The rotor cap is secured tight, and the rotor is loaded on a standard-/heavy-duty cart and transported to the ultracentrifugation unit for spin at $100,000 \times g$ for 70 min (*see* **Notes 7** and **17**).

18. Supernatants are collected in 15–50 mL regular tubes and kept aside (*do not* discard these cell culture supernatants as these can

serve as exosome depleted supernatant, control). The pellet containing exosomes and any contaminants (in ultracentrifugation tubes) is washed/resuspended with ice-cold $1\times$ PBS and spun again at $100,000\times g$ for 70 min (*see* **Note 18**).

19. The resulting pellet of exosomes is considered as exosomal fraction. The pellet of exosomes is reconstituted in $1\times$ PBS. Purified exosome preparations can be used immediately or stored at -80 °C and used later for further analysis (*see* **Note 19**).

3.3 Isolation of Exosomes/EVs from In Vitro Cultures of Mouse or Human N2a Cell Lines or Primary Cultures of Murine Cortical Neurons by Density Gradient (DG-Exo) Method

For isolation of exosomes/EVs by density gradient (DG-Exo) method, please follow **steps (1–4)** from Subheading 3.1 (for mouse/human N2a cells) or **steps (1–14)** from Subheading 3.2 (for isolation of exosomes/EVs from primary cultures of cortical neurons).

1. For isolation of exosome/EVs by DG-Exo ultracentrifugation method, plate neuronal cells at desired density. After desired days of postinfection, cell culture supernatants are collected in 15 to 50 mL centrifugation tubes and spun at $480\times g$ for 10 min; cell pellet is discarded; and the clear supernatant is spun again at $2000\times g$ for 10 min.

2. If desired, the supernatants are filtered using the filtering devices (of 0.1 μm). Also, if higher volumes of supernatants are collected, then samples should be concentrated using the Corning Spin-X UF concentrators or centrifugal filter device with a 5–10 K nominal molecular weight limit (NMWL) and as suggested in Subheadings 3.1 and 3.2 (*see* **Note 24**).

3. Cell culture supernatants were concentrated to ~2–2.5 mL and processed for OptiPrep™ (DG-Exo) isolation as follows. Briefly, discontinuous gradients of 40% (w/v), 20% (w/v), 10%(w/v), and 5% (w/v) solutions of iodixanol are prepared from the stock solution of OptiPrep™ 60% (w/v) of aqueous iodixanol (Axis-Shield PoC, Norway) with 0.25 M sucrose/ 10 mM Tris–HCl, pH 7.5. For the discontinuous gradient, 4 mL of 40%, 20%, and 10% or 3 mL of 5% iodixanol is used (*see* **Note 25**).

4. The concentrated cell culture supernatants are overlaid on to the top of the gradient and centrifuged at $100,000 \times g$ for 18 h at 4 °C (*see* **Note 26**).

5. Six individual fractions of ~3 mL are collected (from top to bottom, as fractions 1–6, with increasing density) and diluted with 5 mL of sterile and chilled $1\times$ PBS. Each fraction is mixed thoroughly with PBS.

6. Fractions are centrifuged again at $100,000 \times g$ for 3 h, 4 °C, and followed by another wash with 5 mL of sterile and chilled $1\times$ PBS at $100,000 \times g$ for 3 h, 4 °C.

7. Pelleted exosomes (as fractions) are resuspended in desired volume of $1\times$ PBS. Purified DG-Exosome preparations can be used immediately or stored at -80 °C and used later for further analysis.

3.4 Isolation of Exosomes/EVs from Mouse Brains by Density Gradient (DG-Exo) Method

This protocol successfully isolates in vivo exosomes as six different fractions using DG-Exo-isolation method from soft tissues such as the brain.

1. C57BL/6 wild-type mice are obtained (from either in-bred laboratory colony or from a commercial vendor) (*see* **Note 1**).

2. Mice are euthanized by an IACUC-approved method. If euthanasia is performed with a WNV-infected mice, approved biosafety facilities are required.

3. After euthanasia, the mice fur is sprayed with ethanol (for sterilization), and the brains are removed and placed on petri plates (60 or 100 mm plates).

4. Mouse brains are sliced (~20 mg) from the center of the tissue. Fresh/frozen brain tissues are sliced lengthwise with razor blade to generate small sections. The tissues are weighed (non-infectious brains can be weighed in a BSL-2-authorized laboratory; however, the infectious brain slices must be weighed inside the biosafety cabinet in an approved ABSL-2 animal facility). Frozen tissues (brain slices) are weighed while partially frozen.

5. Brain slices or tissues are transferred to a tube containing 75 U/mL of collagenase type 3 in Hibernate-E medium (at a ratio of 800 mL per 100 mg of tissue). Each 20 mg tissue is processed as one sample.

6. Tissues are incubated for a total of 20 min in a shaking water bath at 37 °C and are gently mixed by inversion after 5 min, then returned to the incubation, and pipetted for several times after 5–10 min and again kept for the remaining time.

7. After the above incubations, tissues are immediately kept on ice, with the addition of protease and phosphatase inhibitors cocktail. To separate clear supernatants, disassociated tissue is spun at $300 \times g$ for 5 min, followed by transfer into new tubes; the supernatants are centrifuged at $2000 \times g$ for 10 min and again at $10,000 \times g$ for 30 min (all spins are done at 4 °C).

8. If desired, the clear supernatants can be filtered using the filtering devices (of 0.1 μm). Also, if higher volumes of supernatants are collected, then samples should be concentrated using the Corning Spin-X UF concentrators or centrifugal filter device with a 5–10 K nominal molecular weight limit (NMWL) as suggested in Subheading 3.3 (*see* **Note 24**).

9. Supernatants from the brain are concentrated to ~2–2.5 mL and processed for OptiPrep™ (DG-Exos) isolation (as suggested in Subheading 3.3). Briefly, discontinuous gradients of 40% (w/v), 20% (w/v), 10% (w/v), and 5% (w/v) solutions of iodixanol were prepared from the stock solution of OptiPrep™ 60% (w/v) of aqueous iodixanol (Axis-Shield PoC, Norway) with 0.25 M sucrose/10 mM Tris–HCl, pH 7.5. For the discontinuous gradient, 4 mL of 40%, 20%, and 10% or 3 mL of 5% iodixanol is used (*see* **Note 25**).

10. The concentrated supernatants (2–2.5 mL) are overlaid on to the top of the gradient and centrifuged at $100,000 \times g$ for 18 h at 4 °C (*see* **Note 26**).

11. Six individual fractions of ~3 mL are collected (from top to bottom, as fractions 1–6, with increasing density) and diluted with 5 mL of sterile and chilled 1× PBS. Each fraction is mixed thoroughly with PBS.

12. Fractions are centrifuged again at $100,000 \times g$ for 3 h, 4 °C, and followed by another wash with 5 mL of sterile and chilled 1× PBS at $100,000 \times g$ for 3 h, 4 °C.

13. Pelleted exosomes are resuspended in desired volume of 1× PBS. Purified DG-exosome preparations can be used immediately or stored at −80 °C and used later for further analysis.

4   Notes

1. All procedures with infected cultures or infected mice or brain tissues should be carried out at an authorized and regulated BSL-2 laboratory or ABSL-2 facility, respectively. All work should be performed in certified biosafety containment/cabinets and as per the approved institutional biosafety committee (IBC), and/or institutional animal care and use committee (IACUC) protocols. The biosafety cabinets should be designated separately for cell culture or animal procedures. Researchers should follow all the biosafety regulations according to their IBC/IACUC policies. Cell culture work with human cell lines and viruses always requires the use of nitrile gloves and approved PPE, and appropriate procedures should be followed.

2. The CytoOne culture flask comes in different sizes such as T75 (20 sleeves of 5 (100) per case), T150 (8 sleeves of 5 (40) per case), or T225 (5 sleeves of 5 (25) per case). Use bigger flask for isolation of large number of exosomes/EVs. Plate higher densities of cells in bigger flask for better yield of exosomes (The suggested vendor is USA Scientific; CytoOne T25 filter cap TC flask, 10/sleeve, 300/case, Catalog Number CC7682-

4825 are recommended to initiate ATCC cultures). These T25 flask have been treated for better adherence of cells. Use bigger flask as per the requirement to isolate large batch of exosomes/EV.

3. The laboratory virus stocks are generated in either mosquito or Vero cells (for WNV) as host. Briefly, confluent flasks with mosquito or Vero cells are infected with unknown titers of virus and incubated for 7 days (Vero cells) or for 14 days (mosquito cells). The cell culture supernatants (~10–15 mL) are collected in centrifuge tubes and spun at $200 \times g$, 5 min at 4 °C. The clear supernatant is collected in new tube and aliquoted into small O-ring safe tubes with desired volumes. To determine the virus titers, endpoint virus dilution assays or plaque formation assays are performed as described in our previous chapter and other publications [30–33, 35].

4. Due to unavailability of ultracentrifugation rotors to spin higher volumes, one can concentrate the larger volumes of cell culture supernatants from 10 to 50 mL and by using the Corning Spin-X UF concentrators of 5–10 K nominal molecular weight limit (NMWL). For larger volumes of 50 mL, 1–2 concentrators can be used to reduce the volumes. Also, if desired, the supernatants can be filtered using the filtering devices (0.1 μm).

5. For using different rotors in an ultracentrifugation unit, please refer to the Beckman Coulter website for rotor types (https://www.beckman.com/centrifuges/rotors/fixed-angle). Select the tubes and the rotor types according to the manufacturer's recommendations.

6. To know more about the ultracentrifugation tubes, read the manual instruction for use (rotors and tubes, document LR-I-M-24AG.pdf, Beckman Coulter). If infectious cultures are spun, the ultracentrifugation tubes have to be autoclaved or treated with 70% ethanol.

7. A standard or heavy-duty utility cart could be best used to load and transport the ultracentrifugation rotors (these are really heavy) from the biosafety cabinets/cell culture room to the ultracentrifugation unit. Before movement of the cart, make sure that there are no obstacles and traffic in between the cell culture place to the ultracentrifugation unit.

8. The modified RIPA buffer should be supplemented with protease and phosphatase inhibitors cocktail.

9. If mice breeding in-house is desired to generate pregnant female animals, an approved IACUC protocols are required.

10. If pregnant female mice are purchased from a commercial vendor, an approved animal protocol is required. Animals purchased should be no more than E11–E13, as mice has to be

acclimatized for 2–3 days (before use/after transportation) based on the IACUC approval. If 3 days of acclimatization is required, then gestation mice have to be E11/E12 upon arrival so that they can be used as E15/E16 after transport and acclimatization. Some institutions allow for 2 days of acclimatization based on researcher's need and request.

11. All sharp objects must be discarded in appropriate sharp containers.

12. As per the ATCC recommendations, it is important to obtain cell lines (mouse neuro-2a or human N2a cells) from the original resource, due to authentication issues. Authentication of cell lines is highly recommended. Also, ATCC-suggested media and other essential components should be obtained as per their recommendations. Since both mouse and human N2a cells are very prone to morphological changes, we suggest to obtain these cell lines from ATCC and maintain them in the ATCC-recommended complete media. We suggest to use early passages between P2 and P25, as N2a cells tend to morphologically change with higher passage numbers.

13. To obtain a higher yield of EVs, we recommend to plate higher cell densities. This is based on the assumption that higher number of cells would produce and release higher yield of EVs. For reinfection with EVs on naïve cells, plate cells at lower densities as the infection rate is low with EVs when compared to the infection rates with known titers [31].

14. WNV titers can be determined by either virus dilution assay (C6/36 cells) or plaque formation assay (Vero cells) [30–33, 35].

15. For centrifugations at lower speeds ($300 \times g$ and $2000 \times g$), use the regular benchtop centrifuges (for 1.5 to 2.0 mL tubes or 15 to 50 mL tubes). For spinning $10,000 \times g$ for 30 min (for 15 or 50 mL tubes), use the ultracentrifugation unit.

16. It is important to check the ultracentrifugation tubes for cracks (before each use), as this can lead to spill of infectious agents in the rotor (large volumes of 10–15 mL will be hazardous). Any spill in an ultracentrifugation rotor is fairly difficult to clean due to heavy weight and uneasiness in handling.

17. It is important to keep the ultracentrifugation rotor in a refrigerator to allow it to be chilled or cooled as samples need to be at 4° during all spin steps. Samples (in room temperature rotor) can heat up in the ultracentrifugation unit that builds up heat during fast spin.

18. Supernatant collected after first $100,000 \times g$ spin for 70 min is considered as supernatant fraction that is depleted of exosomes. This sample can serve as a control for exosomes pellet

(that comes after the PBS wash step). This sample can be considered as control in later downstream applications such as re-infection of naïve/recipient cells or imaging.

19. Detection of viral loads by QRT-PCR, Western blotting, viral plaque formation assay, hematoxylin and eosin (H&E) staining, immunohistochemical (IHC) method (to determine flavivirus caused neuropathogenesis), and TUNEL assays (to determine neuronal cell death) are discussed in previously published chapter [35].

20. For mice transportation from animal facility to research laboratory by the researchers, special approvals are required from the IACUC.

21. Pregnant female mice are sensitive to anesthetic agents such as isoflurane or acepromazine. Few studies suggest the use of halothane, but conscious cervical dislocation (CD) has been preferred, and this is the best way to euthanize pregnant female mice for collection of embryonic tissues.

22. Instead of NeuroCult chemical disassociation kit, one can use the enzymatic solutions to disassociate the embryonic brain cortices or cortex slices. However, the enzymatic disassociation could be very harsh on soft tissue slices of embryonic brains.

23. Neuronal density is considered very important as plating higher numbers could lead to death of neurons (as neurons should be plated in spacious surfaces to allow the neurite outgrowth and extensions). Also, plating neurons at low densities is not very helpful, as these are nondividing cells.

24. For isolation of exosomes/EVs by density gradient (DG-Exo) method, the supernatants can to be filtered, and the large volumes should get concentrated using the Corning Spin-X UF concentrators or centrifugal filter device with a 5–10 K nominal molecular weight limit (NMWL). The concentration of samples is a required step and not optional.

25. For the discontinuous gradient, we suggest for six fractions. If desired one can make 12 fractions with 4 mL of 40%, 20%, and 10% or 3 mL of 5% iodixanol. However, we suggest for 6 fractions, because if two groups (UI and I) are considered for DG-Exo preparations, 12 fractions can be spun together, as most of the ultracentrifugation rotors will have only 12 places for spin.

26. In the case of DG-Exo method, the first step of ultracentrifugation is for 18 h. It is recommended to start this step in the evening and around 4 PM (as the 18 h spin will be finished by 10 AM in the morning). This also allows an observation window of 1–2 h for successful running of the ultracentrifugation unit.

Acknowledgments

This work was supported by start-up funds from the University of Tennessee, Knoxville, to HS and GN.

References

1. Kimura T, Sasaki M, Okumura E, Sawa H (2010) Flavivirus encephalitis: pathological aspects of mouse and other animal models. Vet Pathol 47(5):806–818

2. Blahove MR, Carter JR (2021) Flavivirus persistence in wildlife populations. Viruses 13(10):2099

3. Chala B, Hamde F (2021) Emerging and re-emerging vector-borne infectious diseases and the challenges for control: a review. Front Public Health 9:715759

4. Chambers TJ, Diamond MS (2003) Pathogenesis of flavivirus encephalitis. Adv Virus Res 60: 273–342

5. Ferraris P, Wichit S, Cordel N, Misse D (2021) Human host genetics and susceptibility to ZIKV infection. Infect Genet Evol 95:105066

6. Kemenesi G, Banyai K (2019) Tick-borne flaviviruses, with a focus on Powassan virus. Clin Microbiol Rev 32(1). https://doi.org/10.1128/CMR.00106-17

7. Kuhn RJ, Dowd KA, Beth Post C, Pierson TC (2015) Shake, rattle, and roll: impact of the dynamics of flavivirus particles on their interactions with the host. Virology 479-480: 508–517

8. Mukhopadhyay S, Kuhn RJ, Rossmann MG (2005) A structural perspective of the flavivirus life cycle. Nat Rev Microbiol 3(1):13–22

9. Pan Y, Cai W, Cheng A, Wang M, Yin Z, Jia R (2022) Flaviviruses: innate immunity, inflammasome activation, inflammatory cell death, and cytokines. Front Immunol 13:829433

10. Schneider WM, Hoffmann HH (2022) Flavivirus-host interactions: an expanding network of proviral and antiviral factors. Curr Opin Virol 52:71–77

11. Vasilakis N, Weaver SC (2017) Flavivirus transmission focusing on Zika. Curr Opin Virol 22: 30–35

12. Xie X, Zeng J (2021) Neuroimmune evasion of Zika virus to facilitate viral pathogenesis. Front Cell Infect Microbiol 11:662447

13. Zhao R, Wnag M, Cao J, Shen J, Zhou X, Wang D et al (2021) Flavivirus: from structure to therapeutics development. Life (Basel) 11(7):615

14. Neal JW (2014) Flaviviruses are neurotropic, but how do they invade the CNS? J Infect 69(3):203–215

15. Sips GJ, Wilschut J, Smit JM (2012) Neuroinvasive flavivirus infections. Rev Med Virol 22(2):69–87

16. Suen WW, Prow NA, Hall RA, Bielefeldt-Ohmann H (2014) Mechanism of West Nile virus neuroinvasion: a critical appraisal. Viruses 6(7):2796–2825

17. Anwar MN, Akhtar R, Abid M, Khan SA, Rehman ZU, Tayyub M et al (2022) The interactions of flaviviruses with cellular receptors: implications for virus entry. Virology 568: 77–85

18. Huang YJ, Higgs S, Horne KM, Vanlandingham DL (2014) Flavivirus-mosquito interactions. Viruses 6(11):4703–4730

19. Jordan TX, Randall G (2016) Flavivirus modulation of cellular metabolism. Curr Opin Virol 19:7–10

20. Klein RS, Diamond MS (2008) Immunological headgear: antiviral immune responses protect against neuroinvasive West Nile virus. Trends Mol Med 14(7):286–294

21. Sultana H, Foellmer HG, Neelakanta G, Oliphant T, Engle M, Ledizet M et al (2009) Fusion loop peptide of the West Nile virus envelope protein is essential for pathogenesis and is recognized by a therapeutic cross-reactive human monoclonal antibody. J Immunol 183(1):650–660

22. Sultana H, Neelakanta G, Foellmmer HG, Montgomery RR, Anderson JF, Koski RA et al (2012) Semaphorin 7A contributes to West Nile virus pathogenesis through TGF-beta1/Smad6 signaling. J Immunol 189(6):3150–3158

23. Paterson R (2005) How West Nile virus crosses the blood-brain barrier. Lancet Neurol 4(1):18

24. Kumar M, Nerurkar VR (2016) In vitro and in vivo blood-brain barrier models to study West Nile virus pathogenesis. Methods Mol Biol 1435:103–113

25. Morrey JD, Olsen AL, Siddharthan V, Motter NE, Wang H, Taro BS et al (2008) Increased blood-brain barrier permeability is not a primary determinant for lethality of West Nile

virus infection in rodents. J Gen Virol 89(Pt 2): 467–473

26. Kobiler D, Lustig S, Gozes Y, Ben-Nathan D, Akov Y (1989) Sodium dodecylsulphate induces a breach in the blood-brain barrier and enables a West Nile virus variant to penetrate into mouse brain. Brain Res 496(1–2): 314–316

27. Verma S, Lo Y, Chapagain M, Lum S, Kumar M, Gurjav U et al (2009) West Nile virus infection modulates human brain microvascular endothelial cells tight junction proteins and cell adhesion molecules: transmigration across the in vitro blood-brain barrier. Virology 385(2):425–433

28. Roe K, Orillo B, Verma S (2014) West Nile virus-induced cell adhesion molecules on human brain microvascular endothelial cells regulate leukocyte adhesion and modulate permeability of the in vitro blood-brain barrier model. PLoS One 9(7):e102598

29. Diamond MS, Klein RS (2004) West Nile virus: crossing the blood-brain barrier. Nat Med 10(12):1294–1295

30. Zhou W, Woodson M, Sherman MB, Neelakanta G, Sultana H (2019) Exosomes mediate Zika virus transmission through SMPD3 neutral sphingomyelinase in cortical neurons. Emerg Microbes Infect 8(1): 307–326

31. Zhou W, Woodson M, Neupane B, Bai F, Sherman MB, Choi KH et al (2018) Exosomes serve as novel modes of tick-borne flavivirus transmission from arthropod to human cells and facilitates dissemination of viral RNA and proteins to the vertebrate neuronal cells. PLoS Pathog 14(1):e1006764

32. Zhou W, Tahir F, Wang JC, Woodson M, Sherman MB, Karim S et al (2020) Discovery of exosomes from tick saliva and salivary glands reveals therapeutic roles for CXCL12 and IL-8 in wound healing at the tick-human skin interface. Front Cell Dev Biol 8:554

33. Vora A, Zhou W, Londodno-Renteria B, Woodson M, Sherman MB, Colpitts TM et al (2018) Arthropod EVs mediate dengue virus transmission through interaction with a tetraspanin domain containing glycoprotein Tsp29Fb. Proc Natl Acad Sci U S A 115(28): E6604–E6613

34. Sultana H, Neelakanta G (2020) Arthropod exosomes as bubbles with message(s) to transmit vector-borne diseases. Curr Opin Insect Sci 40:39–47

35. Sultana H (2016) Examination of West Nile virus neuroinvasion and neuropathogenesis in the central nervous system of a murine model. Methods Mol Biol 1435:83–101

Chapter 10

Protocol to Study West Nile Virus Infection in Brain Slices In Vitro

Parminder J. S. Vig

Abstract

Ex vivo slice cultures of the brain tissue can maintain the cytoarchitecture of the central nervous system (CNS), which allows a thorough understanding of the functions of multiple interconnected cells in a culture system that closely resembles the in vivo environment. Additionally, slice cultures of the brain tissue are advantageous in tracking complex connectivity between neurons and glia both under normal and pathologic conditions, which is not possible in in vitro cell lines. Here, we describe the method of preparing ex vivo slice culture from the mouse cerebellum and the protocol of studying the effects of West Nile virus infection on cerebellar cells.

Key words West Nile, Cerebellum, Slice cultures, Purkinje cells, Calcium-binding proteins, Glia

1 Introduction

The clinical picture of West Nile virus (WNV) infection ranges from a flu-like condition to a more severe neuroinvasive disease, which can lead to death of a patient [1–3]. While neurons are main target of WNV, other cell types especially astrocytes play a significant role in promoting WNV-mediated central nervous system (CNS) damage [1, 3]. However, how intrinsic responses to WNV in specific cell types in different regions of the brain modify CNS pathophysiology is not fully understood.

Here, we describe a protocol for a mouse ex vivo cerebellar slice culture model to study the effects of WNV infection to the CNS. The rationale for using ex vivo system is that slice cultures of the brain tissue have an advantage over the primary cells or cancerous cell lines as they are able to maintain the cytoarchitecture of the CNS, which allows a thorough examination of multiple interconnected cells in a culture system that closely resembles the in vivo environment [4]. Additionally, the slice culture system can be used to compare the early intrinsic CNS response of resident

Fengwei Bai (ed.), *West Nile Virus: Methods and Protocols*,
Methods in Molecular Biology, vol. 2585, https://doi.org/10.1007/978-1-0716-2760-0_10,

cells in different CNS regions during pathologic conditions like WNV infection. Slices are prepared from the cerebellar tissue of 7- to 10-day-old mouse pups. Based on the microgram tissue per slice, 7-day-old cerebellar slice cultures are infected with 1×10^3 to 1×10^7 PFU of WNV. After 12 h of exposure to WNV and 3 days postinfection (dpi) in the normal growth media, the pooled slice cultures are processed either for immunohistochemistry and image analysis, total RNA extraction, and gene expression analysis or for any other studies.

We have examined the effects of WNV infection as early as 3 dpi. Quick and co-workers [4] also reported that at 3 dpi, only 27% of neurons, less than 11% astrocytes, and about 6% microglia were infected with WNV. In another study, Clarke et al. [5] investigated WNV-induced injury to brain slice cultures by determining the levels of lactate dehydrogenase (LDH) in the media of WNV-infected cultures compared to mock-infected controls. No LDH leak was detected at 3 dpi, suggesting negligible injury to the brain slice cultures at 3 dpi [5]. Further, at 3, 5, 7, and 9 dpi, caspase-3 activity assays were used to determine the amount of WNV-induced apoptosis in brain slice cultures. At 3 dpi, the fold increase in caspase-3 activity infected vs mock was not significant. Therefore, 3 dpi to 9 dpi is a good timeline to study the progression of WNV-mediated pathogenesis in specific brain cells infected with WNV.

2 Materials

2.1 Stock Solutions (See Note 1)

1. Opti-MEM (Gibco, Opti-MEM I reduced serum medium, Catalog No. 31985-070).

2. Hank's Balanced Salt Solution (HBSS).

3. Penicillin-streptomycin (10,000 U/mL).

4. Horse serum.

5. 10% D-glucose.

6. Neurobasal medium (Gibco Cat. No 21103-049).

7. B27 (50×) supplement (Gibco Cat. No 17504-044).

8. N2 (100×) supplement (Gibco Cat. No 17502-048).

9. L-Glutamine (100×).

10. D-Glucose, D-(+)-glucose solution 45% in H20 sterile filtered (Sigma-Aldrich Cat. No G8769).

11. Antibiotic/antimycotic (100×).

2.2 Working Solutions

1. B27: Take 50 µL of B27 stock, and make the final volume to 2.5 mL with distilled sterile water.

2. N2: Take 25 µL of N2 stock, and make the final volume to 2.5 mL with sterile distilled water ((1×) working solution).

3. Antimycotic/antibiotic: Take 25 µL of antimycotic/antibiotic stock, and make to 2.5 mL with sterile distilled water ((1×) working solution).

4. L-Glutamine: Take 25 µL of L-glutamine stock, and make to 2.5 mL with sterile distilled water ((1×) working solution).

2.3 Media

2.3.1   Growth Medium

To make 100 mL growth medium, add the following in a sterile flask:

50 mL of Opti-MEM, 25 mL of HBSS, 25 mL of horse serum, and 200 µL of Pen-Strep (stock) (*see* **Note 2**).

2.3.2   Washing Medium

To make 100 mL of washing medium, add the following volumes in a sterile flask:

50 mL of Opti MEM, 50 mL of HBSS, and 4.4 mL of 10% D-glucose (*see* **Note 3**).

2.3.3   Serum-Free Medium

To make 100 mL of serum-free medium, in 93. 89 mL of neurobasal medium (Gibco), add 2 mL of 2% B27 supplement, 1 mL of 1% N2 supplement (Gibco), 1 mL of 1% L-glutamine. 1.11 mL of D-glucose solution, and 1 mL of 1% antibiotic/antimycotic (working solution) (*see* **Note 1**).

2.4 Six-Well Tissue Culture Plates

Six-well tissue culture plates : Use plastic sterile plates with lids from any vendor.

2.5 Millicell Cell Culture Insert (Millipore Sigma, Catalog No. PICMORG50)

Millicell Cell Culture Insert (Millipore Sigma, Catalog No. PIC-MORG50): Needed for growing slice cultures over the growth media.

2.6 Mouse Pups (7–10 Days Old)

Mouse pups (7–10 day old): Cerebral hemispheres or cerebella of these pups are used to prepare tissue culture slices.

2.7 McILwain Tissue Chopper

McILwain Tissue chopper: Required to cut 350 micron thick brain or cerebellar slices.

2.8 CO_2 Incubator

CO_2 Incubator: Needed for incubating slice cultures at 37 °C with 5% CO_2 and 100% humidity.

3 Methods

3.1 Preparation of Brain/Cerebellar Slices

1. 7- to 10-day-old mouse pups (*see* **Notes 4** and **5**).

2. Sacrifice pups and carefully remove the whole brain under sterile conditions in a running hood (*see* **Note 4**).

3. Immerse dissected brains in the washing medium described above, and under a dissecting microscope (in a laminar flow hood), remove meninges (*see* **Note 3**).

4. Dissect out region of interest, for example, the cerebellum, for the preparation of brain slices.

5. Remove meninges-free cerebellum, cut parasagittal slices (350 um thickness) using McILwain Tissue chopper, and separate slices gently into washing medium (Fig. 1) (*see* **Note 6**).

3.2 Culturing of Cerebellar Slices

1. Before opening the Millicell Cell Culture Inserts, prepare the six-well tissue culture plates by adding 1 mL of growth medium just enough to wet the exposed surfaces of the explants, but not to submerge them.

2. Peel off the cover sheet, and use sterilized forceps to remove the insert from the package.

3. Place the insert into the prepared tissue culture plate (*see* **Note 7**).

4. Seed the slices onto the top of moist membrane of the insert (Fig. 2) (*see* **Note 8**).

5. Place tissue culture plates in the CO_2 incubator at 37 °C with 5% CO_2 and 100% humidity.

6. The following day, change the growth medium (1.0 mL) (*see* **Note 9**).

3.3 Incubation with WNV-Containing Media

1. Depending on your study parameters, infect 7-day-old slice cultures growing either in media with serum or serum-free with 1×10^3, 1×10^5, or 1×10^7 PFU of WNV in a BSL-2 or BSL-3 facilities.

2. After 12 h of exposure to WNV, remove media with virus, and add fresh growth media (with serum or serum free) and incubate slices for additional 3–9 days.

3. Pool slices from mock and exposed groups, and process either for immunohistochemistry and image analysis (*see* **Note 10**), total RNA extraction, and gene expression analysis or for any other studies.

4   Notes

1. Prepare all solutions or cell culture media in either non-sterile or sterile high-quality, ultrapure deionized water, respectively, and analytical-grade chemicals and reagents. Store prepared solutions at room temperature or in a refrigerator unless indicated otherwise.

2. Growth or serum-free media can be aliquoted into small volumes and kept at $-20\,^{\circ}C$ up to 1 week. Otherwise, make fresh media every 2–3 days and keep unused sterile media in a refrigerator at $4\,^{\circ}C$.

3. Make fresh media and keep some chilled on ice. Keep dissected out brain tissue in the petri dishes containing the media on ice.

4. Either from the wild-type or transgenic mice expressing genes encoding cell-specific fluorescent protein can be used.

5. An approved IACUC protocol will be required prior to any animal procedures.

6. Use autoclaved 3-M filter papers, which have been cut out into circular disks to cut slices using McILwain Tissue chopper. Slightly wet the paper with wash medium, and gently put the cerebella or other brain tissue in a desired orientation to be cut. Flush slices in an ice-chilled wash medium in a petri dish using an autoclaved glass pipette with a rubber bulb attached to it. Gently shake attached slices to make sure they all singly separated (Fig. 1).

7. Make sure the inserts are positioned correctly. Allow several minutes for the membranes to become moistened.

8. Place slices at the center of the membrane and keep them apart (Fig. 2). Up to three to four cerebellar slices can be placed on each six-well insert.

9. Alternatively, change the growth medium to serum-free medium, 1.0 mL.

10. After finishing your experiment if immunohistochemistry is planned, fix slices (at room temperature) by removing growth or serum-free media, washing for 5 min with 1 mL of sterile PBS and adding 1 mL of 4% freshly prepared paraformaldehyde in PBS for 30 min. Remove fixative and gently wash twice with 1 mL of PBS. Cut membranes, place wet membranes upside down on a glass ++slides, and let the membrane self-dry. Peel off dried membrane and you will see that the intact slice tissue gets transferred to the slide, and then you can proceed to process the slice for immunohistochemistry.

Fig. 1 350-um-thick mouse cerebellar slices cut on McILwain Tissue chopper are suspended in the washing medium (*see* Subheading 2)

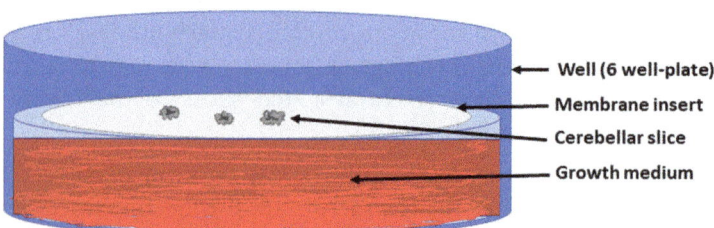

Fig. 2 Schematic showing a single six-well plate with a membrane insert immersed in growth medium (*see* Subheading 2) and cerebellar slices layered on the surface of the membrane

Acknowledgments

This work was supported by funds from the Wilson Foundation, USA.

References

1. Petzold A, Groves M, Leis AA, Scaravilli F et al (2010) Neuronal and glial cerebrospinal fluid protein biomarkers are elevated after West Nile virus infection. Muscle Nerve 41:42–49. https://doi.org/10.1002/mus.21448

2. Blakely PK, Kleinschmidt-DeMasters BK, Tyler KL, Irani DN (2009) Disrupted glutamate transporter expression in the spinal cord with acute flaccid paralysis caused by West Nile virus infection. J Neuropathol Exp Neurol 68:1061–1072. https://doi.org/10.1097/NEN.0b013e3181b8ba14

3. Leis AA, Stokic DS, Petzold A (2012) Glial S100B is elevated in serum across the spectrum

of West Nile virus infection. Muscle Nerve 45: 826–830. https://doi.org/10.1002/mus.23241

4. Quick ED, Leser JS, Clarke P, Tyler KL (2014) Activation of intrinsic immune responses and microglial phagocytosis in an ex vivo spinal cord slice culture model of West Nile virus infection. J Virol 88:13005–13014. https://doi.org/10.1128/JVI.01994-14

5. Clarke P, Leser JS, Quick ED, Dionne KR et al (2014) Death receptor-mediated apoptotic signaling is activated in the brain following infection with West Nile virus in the absence of a peripheral immune response. J Virol 88: 1080–1089. https://doi.org/10.1128/JVI.02944-13

Chapter 11

Studying Virus-Host Interactions with CRISPR Technology

Rong Zhang

Abstract

The mosquito-borne West Nile virus (WNV) poses a great threat to public health as no vaccine or specific antiviral treatment is available. Exploring virus-host interactions, specifically host factors that are required for virus infection, is important for better understanding the biology, pathogenesis, and transmission of WNV. Such essential host factors may also represent antiviral targets. The development of CRISPR technology has provided a powerful and convenient tool to perturbate host gene expression, allowing for unbiased, genome-wide screens of host factors for virus infection. Here we describe the necessary steps for performing a CRISPR knockout screen, which can also be applied to other viruses, to identify host factors critical for WNV infection.

Key words West Nile virus, CRISPR, Genome-wide screen, Host factor, Virus-host interactions

1 Introduction

West Nile virus (WNV) is a mosquito-borne encephalitic virus that belongs to the *Flavivirus* genus of the family *Flaviviridae* [1]. Since its discovery in Uganda in 1937, WNV has spread to multiple continents, including Europe, Asia, and North America [2], making disease control and prevention difficult, especially as no vaccine or antiviral agent is available. WNV is an enveloped virus with a single-stranded, positive-sense RNA genome approximately 11 kb in length [3]. The genome contains one open reading frame, which is translated as a single polyprotein that is post- and co-translationally cleaved by both viral and host proteases, resulting in three structural and seven nonstructural proteins. Noncoding regions are located at the 5' and 3' ends of the genome and form stem-loop structures that aid in replication, transcription, translation, and packaging [3]. Although these RNA structures and proteins have diverse functions, the compact genome and limited number of proteins encoded require the use of host factors to facilitate infection.

Fengwei Bai (ed.), *West Nile Virus: Methods and Protocols*,
Methods in Molecular Biology, vol. 2585, https://doi.org/10.1007/978-1-0716-2760-0_11,

WNV can infect a wide variety of species, including mosquitoes, birds, humans, and other vertebrates [1]. The transmission between different hosts suggests that WNV can utilize conserved host factors to replicate. Thus, exploring the interplay between host factors and virus-encoded elements is of great help in understanding the basic biology, pathogenesis, and transmission of this virus. Moreover, novel specific or conserved host factors that are identified with defined mechanisms of action could be appealing antiviral targets for drug discovery. Targeting host factors that are conserved and essential for virus infection would be less likely to acquire drug resistance.

To identify host factors associated with virus infection, various strategies have been utilized, including the yeast two-hybrid system, transcriptomic analysis, affinity-purification coupled to mass spectrometry, and cDNA library screens [4–7]. In addition, direct perturbation of host gene expression by RNA interference (RNAi), gene trap mutagenesis, and especially the nuclease-based CRISPR (clustered regularly interspaced short palindromic repeat) technology have significantly advanced the identification of virus-related host factors [8–10]. Despite the ease with which screens can be conducted using RNAi short hairpin RNA (shRNA), the high off-target effects and incomplete knockdown of transcript levels hamper their use in interrogating the biological functions of host genes [11]. Although gene trap mutagenesis can introduce complete loss of gene function and has contributed to identifying essential host entry factors, such as Niemann-Pick C1 protein (NPC1) for Ebola virus [12], lysosome-associated membrane protein-1 (LAMP1) for Lassa virus [13], and protocadherin 1 (PCHD1) for hantavirus [14], specific haploid cells have to be used, limiting the broad application of this technique to viruses that do not replicate in these cells.

The recently developed CRISPR-Cas9 gene editing tool utilizes a single-guide RNA (sgRNA) with a Cas9 nuclease to cleave host genomic DNA, resulting in insertions or deletions (indels) that perturbate host gene expression [15, 16]. Fusion of a sgRNA with a nuclease-dead Cas9 with transcription activators or repressors can overexpress or inhibit host gene expression, respectively [17, 18]. Given that the editing is rapid, efficient, and cost-effective and that the off-target effect is very low compared to RNAi/shRNA [19, 20], CRISPR technology has been widely employed to study the diverse biological functions of genes both in vitro and in vivo. Additionally, the relative ease of generating a library of sgRNAs targeting a whole genome has allowed for high-throughput screens to map specific host genes to a phenotype of interest in an unbiased manner [21–23].

Here we provide a detailed protocol to conduct a genome-wide CRISPR knockout screen using the ready-made human sgRNA library to identify host factors that promote WNV infection. The

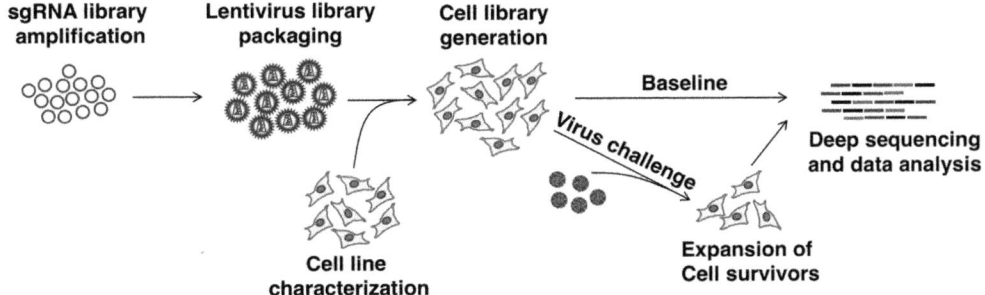

Fig. 1 Schematic of a CRISPR screen for proviral host factors. The ready-made or custom sgRNA plasmid library is amplified in bacteria. After plasmid purification and packaging, the lentivirus library is transduced into target cells to generate the cell library for viral challenge. The control cells and expanded virus-resistant cells are subjected to genomic DNA extraction and PCR amplification of the sgRNA cassettes followed by next-generation deep sequencing and data analysis

steps for sgRNA library amplification and packaging, cell line characterization, cell library preparation, virus challenge, screening, DNA sequencing, and data analysis are described (Fig. 1). This protocol can also be applied to screen host factors required for other viruses in the cell line of choice.

2   Materials

2.1 Cell Culture

1. Dulbecco's Modified Eagle Medium (DMEM, high glucose) (Thermo Fisher 11965118).

2. Fetal bovine serum (FBS).

3. 1 M HEPES buffer solution.

4. 10,000 U/mL penicillin-streptomycin.

5. 10% FBS culture medium: 10% FBS in DMEM, 10 mM HEPES, 1× penicillin-streptomycin.

6. 293FT cells (Thermo Fisher R70007).

7. Human embryonic kidney (HEK) 293 T cells (ATCC CRL-3216).

8. Class II, type A2 biosafety cabinet.

9. 0.05% trypsin-EDTA.

10. Cell culture incubator, 5% CO_2, 37 °C.

11. Serological pipette, 5, 10, and 25 mL.

12. Flasks (T25, T175), tissue culture treated.

13. 12-well plates, tissue culture treated.

2.2 Library Amplification and Packaging

1. pMD2.G (Addgene 12259).
2. psPAX2 (Addgene 12260).
3. lentiCas9-Blast (Addgene 52962).
4. GeCKO v2 pooled library (Addgene 1000000049).
5. Endura ElectroCompetent Cells (Lucigen 60242).
6. Endo-Free Plasmid Maxi Kit (Qiagen 12362).
7. MicroPulser cuvette, 0.1 cm gap (Bio-Rad 1652083).
8. Gene Pulser Xcell Microbial System (Bio-Rad).
9. Recovery medium (Endura 80026).
10. FuGENE HD transfection reagent (Promega E2311).
11. Opti-MEM I Reduced-Serum Medium (Thermo Fisher 31985).
12. LB agar.
13. LB broth.
14. NanoDrop One (Thermo Fisher).
15. Shaking incubator.

2.3 Cell Library Preparation

1. Puromycin dihydrochloride (Thermo Fisher A1113803).
2. Blasticidin S HCl (Thermo Fisher A1113903).
3. Polybrene transfection reagent (Millipore TR-1003-G).
4. Benchtop microcentrifuge.
5. Countess II automated cell counter.

2.4 WNV Infection and CRISPR Screen

1. WNV stock.
2. Biosafety level 2/3 laboratory.
3. 10% Bleach.

2.5 Genomic DNA Extraction and Sequencing

1. QIAamp DNA Blood Maxi Kit (Qiagen 51192).
2. PfuUltra II Fusion HS DNA polymerase (Agilent 600670).
3. QIAquick Gel Extraction Kit (Qiagen 28706).
4. MAGeCK software.
5. FASTX-Toolkit (http://hannonlab.cshl.edu/fastx_toolkit/).
6. Cutadapt 1.8.1.
7. HiSeq 3000 or similar platform (Illumina).
8. Primers for amplifying and sequencing the sgRNA cassettes:

   Primers for the first round of PCR

F1:   5′- AATGGACTATCATATGCTTACCGTAACTTGAAAG TATTTCG-3′.

R1:   5'- CTTTAGTTTGTATGTCTGTTGCTATTATGTCTAC TATTCTTTCC-3'.

Primers for the second round of PCR.

F2: 5'- AATGATACGGCGACCACCGAGATCTACACTCTTTC CCTACACGACGCTCTTCCGATCT (1-9 bp stagger) TCT TGTGGAAAGGACGAAACACCG-3'.

R2:5'- CAAGCAGAAGACGGCATACGAGAT (7–8 bp barcode) GTGACTGGAGTTCAGACGTGTGCTCTTCCGATCTAC TATTCTTTCCCCTGCACTGT-3'.

3   Methods

3.1   sgRNA Library Amplification

1. Dilute the genome-wide human GeCKO v2 sgRNA plasmid library to 50 ng/μL in water or TE buffer, and pre-chill on ice (*see* **Note 1**).

2. Equilibrate the Endura recovery medium to room temperature.

3. Warm the LB agar plates to 37 °C in an incubator.

4. Pre-chill four MicroPulser cuvettes and 1.5-mL microcentrifuge tubes on ice.

5. Thaw the Endura electrocompetent cells on ice completely and aliquot 25 μL per tube to the chilled microcentrifuge tubes on ice.

6. Add 2 μL of the sgRNA plasmid library to the chilled competent cells, and stir briefly with a pipette tip without introducing air bubbles (*see* **Note 2**).

7. Transfer the cell/plasmid mixture into a chilled cuvette carefully without introducing air bubbles.

8. Place the cuvette on the Gene Pulser Xcell Microbial System and perform the electroporation according to the manufacturer's suggested conditions.

9. Add 975 μL of recovery medium to the cuvette and transfer to a culture tube with an additional 1 mL of recovery medium.

10. Repeat the electroporation above three additional times and place the tubes in a shaking incubator at 200 rpm for 1 h at 37 °C.

11. Pool and mix all approximately 8 mL of electroporated cells. To titrate the library, add 10 μL of the pooled, recovered cells to 990 μL of recovery medium, and plate 20 μL onto a pre-warmed 10-cm LB agar plate. Plate the remaining volume of pooled, recovered cells across 20 pre-warmed 10-cm LB agar plates (~400 μL/plate).

12. Incubate all plates at 32 °C for 14 h (*see* **Note 3**).

13. Calculate the transformation efficiency by multiplying the number of colonies on the titration plate by 40,000 for the total number of colonies on all plates. Proceed if the number represents a library coverage of at least 50-fold (i.e., at least 3×10^6 colonies for each sub-library) (*see* **Note 4**).

14. Add 500 µL of LB broth to each 10-cm LB agar plate and scrape the colonies with a spreader into a pre-weighed tube. Repeat once.

15. Centrifuge the tubes at $2500 \times g$ for 10 min at 4 °C to pellet the cells. Weigh the cell pellet and purify the amplified plasmid library using the Endo-Free Plasmid Maxi Kit according to the manufacturer's instructions (*see* **Note 5**).

16. Measure the plasmid concentration with a NanoDrop.

3.2 sgRNA Library Packaging

Lentivirus packaging and transduction should be performed in a class II, type A2 biosafety cabinet in a biosafety level 2 laboratory. Personal protection equipment is required. All the waste produced should be bleached or autoclaved before disposal.

1. For every 30 mL of lentivirus being produced, seed 1.5×10^7 293FT cells per T175 flask 1 day prior to transfection (*see* **Note 6**).

2. When cells reach around 90% confluence, prepare the transfection mixture as follows to package the sgRNA lentivirus library: For each T175 flask, add 7.5 µg pMD2.G, 12.5 µg psPAX2, and 15 µg sgRNA plasmid library into 3.5 mL of Opti-MEM; mix well. Then add 105 µL of FuGENE HD transfection reagent to the plasmid mixture and mix well.

3. Incubate the transfection mixture at room temperature for 15 min, and add to the 293FT cells.

4. After 6–8 h, replace the medium with 30 mL of fresh 10% FBS culture medium.

5. At 48 h post transfection, harvest the culture supernatant, and spin at $2500 \times g$ for 15 min to remove the cell debris. Filter the debris-free supernatant through a 0.45-µm filter, aliquot, and store at −80 °C (*see* **Note 7**).

3.3 Characterization of the Cell Line for Screening

1. To conduct the cell survival-based screen for WNV, pick (a) cell line(s) highly permissive to virus infection. Here, we use 293T cells as an example as they are commonly used in the laboratory and most of the cells die 1 week after virus challenge (*see* **Note 8**).

2. Seed cells in 12-well plates and infect with WNV from low to high multiplicity of infection (MOI) when cells reach confluency. Check the cytopathic effect daily, and choose (a) cell

line(s) that can be killed largely by the virus infection in around 1 week. For 293T cells, infect at an MOI of 1 can kill most of the cells in less than a week.

3. Optimize the concentration of puromycin and blasticidin on the chosen cell line for the screening. Seed cells in 12-well plates at the density required to reach confluency after 3–4 days. Add puromycin or blasticidin to each well at concentrations ranging from 0 to 20 µg/mL. Observe the cells daily, and select the lowest concentration of each drug that kills all the cells in 3–4 days. For 293 T cells, the optimal concentration of puromycin and blasticidin is 2 µg/mL and 12 µg/mL, respectively.

4. To increase transduction efficiency, polybrene is often added. Assess the potential toxicity of polybrene on (the) chosen cell line(s) as done for puromycin and blasticidin above. Typically, the range of concentration is 2–8 µg/mL. For 293T cells, 8 µg/mL of polybrene can be used.

3.4 Cell Library Preparation

1. For the two-component screening system, generate the cell line stably expressing Cas9 first. Package the lentivector lentiCas9-Blast as done for the sgRNA library in Subheading 3.2, and transduce the chosen cell line in a 12-well plate with the packaged lentivirus. In parallel, seed an additional well to serve as the nontransduced control cells. Select the transduced and nontransduced control cells with blasticidin (i.e., 12 µg/mL for 293 T cells) for around 4 days or until the time that the nontransduced control cells have all died. The Cas9 cell line can be expanded and used to generate the cell library. To increase the performance of the screen, the clonal cell line may instead be used (*see* **Note 9**).

2. To reduce the possibility that one cell has multiple sgRNAs integrated, titrate the dose of lentivirus library used for the transduction so that the MOI is approximately 0.3 or lower (*see* **Note 10**). Seed the chosen cell line stably expressing the Cas9 (i.e., 2×10^5 of 293 T-Cas9 cells) in 1 mL of culture medium per well in 12-well plates 24 h prior to transduction with the sgRNA library. Make sure to seed an extra well as a negative (i.e., nontransduced) control. Count the cell number at the time of transduction when cells reach 30–50% confluence.

3. Add different volumes of the packaged lentivirus library from 0 to 200 µL to each well, with a total volume of 1 mL/well. Add the polybrene reagent to each well (at a final concentration of 8 µg/mL for 293T cells).

4. Spinfect the cells by centrifuging the plates at $1000 \times g$ for 1 h at 32 °C (*see* **Note 11**).

5. After 24 h, use 0.05% trypsin-EDTA to trypsinize, and transfer the cells from each well to a T25 flask. Add the optimized concentration of puromycin in Subheading 3.3 (at a final concentration of 2 μg/mL for 293 T cells) (*see* **Note 12**).

6. After selection with puromycin for around 3 days (after all the nontransduced, puromycin-treated cells have died), count the number of cells, and identify the lentivirus dose required for ~30% cell survival (the number of cells in the nontransduced, untreated flask divided by the number of cells in the transduced, puromycin-treated flask). Calculate the number of cells required to represent the entire sgRNA library at more than 500-fold coverage. For example, using the GeCKO v2 library A or B and 293 T cells as an example, the calculation for how many total cells are required would be ~60,000 sgRNAs × 500 (for 500-fold coverage), divided by 0.3 (for ~30% cell survival), and then this amount is multiplied by the quotient of cells seeded (2×10^5) and divided by the cell count per well at the time of transduction. Dividing this quotient by 12 will yield the number of 12-well plates to seed.

7. Prepare the required number of 12-well plates of Cas9 stable-expressing cells, and spinfect at $1000 \times g$ for 1 h at 32 °C with the lentivirus library at the dose determined above in the presence of polybrene. Also, prepare the nontransduced control (*see* **Note 13**).

8. At 24-h post transduction, trypsinize and transfer the cells into T175 flasks, and select with the appropriate dose of puromycin (2 μg/mL for 293T cells) (*see* **Note 14**).

9. When the cells reach ~100% confluence, passage the number of cells that represent at least 500-fold coverage of the sgRNA library as determined in **step 6**. After transduction for 7 days, the cell library is ready for the screen without further selection with puromycin. The nontransduced cells treated with puromycin should all be dead.

3.5 WNV Infection and Screen of the Cell Library

The cell culture infection with WNV should be conducted in a biosafety level 2/3 laboratory in strict accordance with safety regulations and best practices.

1. To conduct the screen for proviral host factors, seed T175 flasks with the cell library generated above 24 h prior to the virus infection for each technical replicate. For 293T cells, seed about six T175 flasks with 1.5×10^7 cells per T175. Maintain at least 3.5×10^7 cells as uninfected control in parallel.

2. When cells reach ~100% confluence, infect with the optimized MOI of WNV that can kill the cells in around 1 week. Check the virus-induced cytopathic effect daily. In parallel, passage the uninfected control cells as they reach confluency.

3. About 1 week later, when the majority of cells are dead, replace the media (10% FBS culture medium), and let the residual cells grow up.

4. Once cell colonies form, trypsinize the cells, and pool and transfer to two to three new flasks until they are near confluency.

5. Rechallenge these cells with WNV as done above in **steps 2** and **3** (*see* **Note 15**).

6. Trypsinize and combine the cell colonies for a given technical replicate, and seed into 1–2 new flasks.

7. Harvest 3.5-5 $\times$ 10^7 uninfected control cells and ~ 3 $\times$ 10^7 of the cells that survived infection for genomic DNA extraction.

3.6 Genomic DNA Extraction and Amplicon Preparation

1. Extract genomic DNA using the QIAamp DNA Blood Maxi Kit according to the manufacturer's instructions (*see* **Note 16**).

2. Combine the eluted genomic DNA for each technical replicate, and measure the concentration with a NanoDrop.

3. To prepare the PCR amplicons for sequencing, PCR is performed in two steps (*see* **Note 17**). For the first PCR, use 2.5 µg of genomic DNA per 50-µL reaction.

4. Prepare each PCR mixture as follows: 2.5 µg of genomic DNA, 2 µL of 10-µM primer F1, 2 µL of 10-µM primer R1, 5 µL of 10× PfuUltra II reaction buffer, 2 µL of 10-mM dNTP, 1 µL of PfuUltra II Fusion HS DNA polymerase; add dH$_2$O to make a 50-µL reaction volume. Amplify the reactions using the following program: 95 °C 2 min; 18 cycles of 95 °C for 20 s, 58 °C for 20 s, and 72 °C for 20 s and lastly 72 °C for 3 min. Combine the resulting PCR products for each replicate.

5. Conduct the second round of PCR in a 50-µL reaction volume using 2.5 µL per reaction of the first PCR product. Set up the PCR as done above in **step 4**, except that the primer pair F2/R2 is used (*see* **Note 18**). Pool the PCR product for each replicate.

6. Run the PCR product on a 2% agarose gel. Excise the ~350 bp target band, and purify the DNA using the QIAquick Gel Extraction Kit according to the manufacturer's instructions (*see* **Note 19**).

7. Measure the amplicon concentration with a NanoDrop or Qubit, and submit for sequencing on an Illumina next-generation sequencing platform such as HiSeq or NextSeq. A read depth of 500-fold sgRNA library coverage may be required (*see* **Note 20**).

3.7 Data Analysis

1. Demultiplex the sequencing data based on the barcode sequences used for each sample.

2. Remove the adaptors and unrelated sequences using the FASTX-Toolkit (http://hannonlab.cshl.edu/fastx_toolkit/) and cutadapt 1.8.1.

3. Map, count, and analyze the reads using the MAGeCK software. A gene list with ranks and sgRNA scores is given based on the positive or negative screening strategy. The reads can also be analyzed using the MAGeCK-associated tools, such as MAGeCK-VISPR and MAGeCKFlute (*see* **Note 21**). Similar data analysis strategy can be applied to other genome-wide CRISPR libraries available (*see* **Notes 22** and **23**).

4   Notes

1. The genome-wide human GeCKO v2 sgRNA library provided by Addgene is at the concentration of around 50 ng/μl, which is ready for transformation without dilution.

2. When mixing the plasmid library with competent cells or transferring the mixture of plasmid and cells to the cuvette, avoid introducing any bubbles as these can lead to arcing during the electroporation, resulting in cell death and low transformation efficiency.

3. Growing the plates at low temperature may reduce recombination. Also, avoid incubating the plates too long as this may introduce bias in the sgRNA library distribution due to differences in colony growth rate.

4. The human GeCKO v2 sgRNA library has been divided into two sub-libraries. Each sub-library has three sgRNAs per gene for a total number of $\sim 6 \times 10^4$ sgRNAs. Make sure that there is at least 50-fold coverage of the sgRNA library.

5. Approximately 1 g of wet cell pellet can be processed by one maxi column. Overloading the column will decrease the plasmid purification efficiency.

6. The 293FT cells are easier to transfect and give higher lentiviral titers than 293 T cells. They also grow faster than 293T cells. However, 293T cells can still be used to package the lentivirus. If the titer is low, use ultracentrifugation to concentrate the virus.

7. The virus stock can be stored at 4 °C for less than 1 week. Try to reduce the number of freeze-thaw cycles. Aliquot some vials to use for titrating the MOI for library generation.

8. To perform a positive-selection screen to identify host genes that when knocked out result in cells becoming resistant to

infection, it is critical to select an appropriate cell line. For WNV, a number of cell lines are susceptible to infection. Choose lines that are highly permissive and die after virus infection.

9. Generating a clonal cell line by limiting dilution and using the cell line with the high Cas9 expression as validated by western blot can increase screening performance.

10. Transduction at a higher MOI such as 0.5 is fine but will increase the false-positive rate and require more work to validate the gene hits.

11. Spinfecting the cells with lentivirus can increase transduction efficiency. A longer centrifugation time of 2 h is acceptable. If some cells do not tolerate spinfection, just perform the infection without centrifugation.

12. Make sure the untreated cells are not growing to confluence as the proliferating rate is different compared to drug-treated cells with low cell confluence. Passage both treated and untreated cells when the untreated cells are near confluence.

13. The cell library should be generated under the same conditions as optimized previously. To increase reliability, use the same culture plates, culture media volume, cell density, timing of transduction with lentivirus, etc.

14. It is recommended to determine the transduction efficiency of the cell library. Pool the cells and seed into two T25 flasks, one with and one without puromycin treatment. After selection, count the number of cells and calculate the ratio of treated to untreated cells.

15. Three rounds of virus challenge can be done to enrich the resistant cells even more and identify critically important genes required for virus infection.

16. Do not overload the columns with the genomic DNA. No clumps should be present in the solution.

17. The amplicons can also be prepared by one-step PCR. The primer pairs for one-step PCR will need to contain not only the binding sequences for the sgRNA cassette but also the adaptor and barcode sequences so that the PCR products can be directly subjected for deep sequencing.

18. To increase the sequence complexity of the sgRNA amplicons for deep sequencing, the forward primers contain a 1–9 bp length of stagger sequence.

19. A nonspecific band above the ~350 bp target band may be visible. Run the gel at low voltage to separate the bands and excise the right one for DNA purification.

20. For positive screening of infected samples, since the survival cells are highly enriched, the sequencing coverage can be reduced.

21. In addition to MAGeCK, other tools based on different algorithms can be applied, such as STARS (https://portals.broadinstitute.org/gpp/public/software/index) [24].

22. A number of existing, ready-to-use CRISPR knockout libraries are available from the nonprofit organization Addgene or other commercial companies and can be employed to conduct the positive screen to identify host dependency factors for virus infection.

23. Compared to the wild-type Cas9-based knockout screen to identify host factors required for virus infection, dCas9 (nuclease-dead Cas9) fused with transcription activators or repressors to up- or downregulate gene expression can be employed to identify antiviral or proviral host factors, respectively [25–27]. The screening is performed similar to what is described in this protocol.

References

1. Colpitts TM, Conway MJ, Montgomery RR, Fikrig E (2012) West Nile virus: biology, transmission, and human infection. Clin Microbiol Rev 25(4):635–648. https://doi.org/10.1128/CMR.00045-12

2. Chancey C, Grinev A, Volkova E, Rios M (2015) The global ecology and epidemiology of West Nile virus. Biomed Res Int 2015:376230. https://doi.org/10.1155/2015/376230

3. Chambers TJ, Hahn CS, Galler R, Rice CM (1990) Flavivirus genome organization, expression, and replication. Annu Rev Microbiol 44:649–688. https://doi.org/10.1146/annurev.mi.44.100190.003245

4. Uetz P, Hughes RE (2000) Systematic and large-scale two-hybrid screens. Curr Opin Microbiol 3(3):303–308. https://doi.org/10.1016/s1369-5274(00)00094-1

5. Ma-Lauer Y, Lei J, Hilgenfeld R, von Brunn A (2012) Virus-host interactomes – antiviral drug discovery. Curr Opin Virol 2(5):614–621. https://doi.org/10.1016/j.coviro.2012.09.003

6. Ploss A, Evans MJ, Gaysinskaya VA, Panis M, You H, de Jong YP, Rice CM (2009) Human occludin is a hepatitis C virus entry factor required for infection of mouse cells. Nature 457(7231):882–886. https://doi.org/10.1038/nature07684

7. DeFilippis V, Raggo C, Moses A, Fruh K (2003) Functional genomics in virology and antiviral drug discovery. Trends Biotechnol 21(10):452–457. https://doi.org/10.1016/S0167-7799(03)00207-5

8. Carette JE, Guimaraes CP, Varadarajan M, Park AS, Wuethrich I, Godarova A, Kotecki M, Cochran BH, Spooner E, Ploegh HL, Brummelkamp TR (2009) Haploid genetic screens in human cells identify host factors used by pathogens. Science 326(5957):1231–1235. https://doi.org/10.1126/science.1178955

9. Puschnik AS, Majzoub K, Ooi YS, Carette JE (2017) A CRISPR toolbox to study virus-host interactions. Nat Rev Microbiol 15(6):351–364. https://doi.org/10.1038/nrmicro.2017.29

10. Paddison PJ, Silva JM, Conklin DS, Schlabach M, Li M, Aruleba S, Balija V, O'Shaughnessy A, Gnoj L, Scobie K, Chang K, Westbrook T, Cleary M, Sachidanandam R, McCombie WR, Elledge SJ, Hannon GJ (2004) A resource for large-scale RNA-interference-based screens in mammals. Nature 428(6981):427–431. https://doi.org/10.1038/nature02370

11. Jackson AL, Bartz SR, Schelter J, Kobayashi SV, Burchard J, Mao M, Li B, Cavet G, Linsley PS (2003) Expression profiling reveals off-target gene regulation by RNAi. Nat

Biotechnol 21(6):635–637. https://doi.org/10.1038/nbt831

12. Carette JE, Raaben M, Wong AC, Herbert AS, Obernosterer G, Mulherkar N, Kuehne AI, Kranzusch PJ, Griffin AM, Ruthel G, Dal Cin P, Dye JM, Whelan SP, Chandran K, Brummelkamp TR (2011) Ebola virus entry requires the cholesterol transporter Niemann-Pick C1. Nature 477(7364):340–343. https://doi.org/10.1038/nature10348

13. Jae LT, Raaben M, Herbert AS, Kuehne AI, Wirchnianski AS, Soh TK, Stubbs SH, Janssen H, Damme M, Saftig P, Whelan SP, Dye JM, Brummelkamp TR (2014) Virus entry. Lassa virus entry requires a trigger-induced receptor switch. Science 344(6191):1506–1510. https://doi.org/10.1126/science.1252480

14. Jangra RK, Herbert AS, Li R, Jae LT, Kleinfelter LM, Slough MM, Barker SL, Guardado-Calvo P, Roman-Sosa G, Dieterle ME, Kuehne AI, Muena NA, Wirchnianski AS, Nyakatura EK, Fels JM, Ng M, Mittler E, Pan J, Bharrhan S, Wec AZ, Lai JR, Sidhu SS, Tischler ND, Rey FA, Moffat J, Brummelkamp TR, Wang Z, Dye JM, Chandran K (2018) Protocadherin-1 is essential for cell entry by New World hantaviruses. Nature 563(7732):559–563. https://doi.org/10.1038/s41586-018-0702-1

15. Mali P, Yang L, Esvelt KM, Aach J, Guell M, DiCarlo JE, Norville JE, Church GM (2013) RNA-guided human genome engineering via Cas9. Science 339(6121):823–826. https://doi.org/10.1126/science.1232033

16. Cong L, Ran FA, Cox D, Lin S, Barretto R, Habib N, Hsu PD, Wu X, Jiang W, Marraffini LA, Zhang F (2013) Multiplex genome engineering using CRISPR/Cas systems. Science 339(6121):819–823. https://doi.org/10.1126/science.1231143

17. Konermann S, Brigham MD, Trevino AE, Joung J, Abudayyeh OO, Barcena C, Hsu PD, Habib N, Gootenberg JS, Nishimasu H, Nureki O, Zhang F (2014) Genome-scale transcriptional activation by an engineered CRISPR-Cas9 complex. Nature. https://doi.org/10.1038/nature14136

18. Gilbert LA, Horlbeck MA, Adamson B, Villalta JE, Chen Y, Whitehead EH, Guimaraes C, Panning B, Ploegh HL, Bassik MC, Qi LS, Kampmann M, Weissman JS (2014) Genome-scale CRISPR-mediated control of gene repression and activation. Cell 159(3):647–661. https://doi.org/10.1016/j.cell.2014.09.029

19. Morgens DW, Deans RM, Li A, Bassik MC (2016) Systematic comparison of CRISPR/Cas9 and RNAi screens for essential genes. Nat Biotechnol 34(6):634–636. https://doi.org/10.1038/nbt.3567

20. Evers B, Jastrzebski K, Heijmans JP, Grernrum W, Beijersbergen RL, Bernards R (2016) CRISPR knockout screening outperforms shRNA and CRISPRi in identifying essential genes. Nat Biotechnol 34(6):631–633. https://doi.org/10.1038/nbt.3536

21. Wang T, Wei JJ, Sabatini DM, Lander ES (2014) Genetic screens in human cells using the CRISPR-Cas9 system. Science 343(6166):80–84. https://doi.org/10.1126/science.1246981

22. Shalem O, Sanjana NE, Hartenian E, Shi X, Scott DA, Mikkelsen TS, Heckl D, Ebert BL, Root DE, Doench JG, Zhang F (2014) Genome-scale CRISPR-Cas9 knockout screening in human cells. Science 343(6166):84–87. https://doi.org/10.1126/science.1247005

23. Koike-Yusa H, Li Y, Tan EP, Velasco-Herrera Mdel C, Yusa K (2014) Genome-wide recessive genetic screening in mammalian cells with a lentiviral CRISPR-guide RNA library. Nat Biotechnol 32(3):267–273. https://doi.org/10.1038/nbt.2800

24. Doench JG, Fusi N, Sullender M, Hegde M, Vaimberg EW, Donovan KF, Smith I, Tothova Z, Wilen C, Orchard R, Virgin HW, Listgarten J, Root DE (2016) Optimized sgRNA design to maximize activity and minimize off-target effects of CRISPR-Cas9. Nat Biotechnol 34(2):184–191. https://doi.org/10.1038/nbt.3437

25. Orchard RC, Sullender ME, Dunlap BF, Balce DR, Doench JG, Virgin HW (2019) Identification of antinorovirus genes in human cells using genome-wide CRISPR activation screening. J Virol 93(1). https://doi.org/10.1128/JVI.01324-18

26. Dukhovny A, Lamkiewicz K, Chen Q, Fricke M, Jabrane-Ferrat N, Marz M, Jung JU, Sklan EH (2019) A CRISPR activation screen identifies genes that protect against Zika virus infection. J Virol 93(16). https://doi.org/10.1128/JVI.00211-19

27. Heaton BE, Kennedy EM, Dumm RE, Harding AT, Sacco MT, Sachs D, Heaton NS (2017) A CRISPR activation screen identifies a Pan-avian influenza virus inhibitory host factor. Cell Rep 20(7):1503–1512. https://doi.org/10.1016/j.celrep.2017.07.060

Chapter 12

Protocol of Detection of West Nile Virus in Clinical Samples

Hephzibah Nwanosike, Freedom M. Green, Kristy O. Murray, Jill E. Weatherhead, and Shannon E. Ronca

Abstract

West Nile virus (WNV) is one of the leading causes of arboviral encephalitis in the United States but is often underdiagnosed. Despite the wide breadth of WNV-induced clinical disease syndromes, many of the symptoms associated with WNV are nonspecific at the time of presentation; thus, choosing the right diagnostic tool is essential to not only understand the true burden of disease but also provide pathogen-directed interventions for WNV-infected patients. In this chapter, we briefly discuss the three most common types of diagnostic methods for WNV in human clinical samples: nucleic acid detection, enzyme-linked immunoassay (ELISA), and plaque reduction neutralization test (PRNT) and present the method for PRNT.

Key words West Nile virus, Flavivirus, Diagnostics, PRNT

1 Introduction

West Nile virus (WNV) is spread through the bite of an infected mosquito. It is one of the leading causes of arboviral infection in North America, with an estimated seven million people infected in the United States alone [1]. Transmission from mosquito to humans commonly occurs during summer-early fall when mosquito activity is highest. Most individuals infected with WNV (80%) have subclinical disease, while others present with a mild to moderate acute febrile illness that progresses to severe neurological disease. Of the ~20% of WNV-infected individuals that develop West Nile fever, ~1% will develop encephalitis, meningitis, and/or acute flaccid paralysis [1, 2]. These infections can leave individuals with long-term complications in up to 50% of infected individuals [3–9], particularly in the elderly and immunocompromised. Due to the long-term health impact of WNV, it is essential that healthcare providers identify these infections. Diagnosis of WNV based on clinical presentation alone is challenging due to the similarity of

Fengwei Bai (ed.), *West Nile Virus: Methods and Protocols*,
Methods in Molecular Biology, vol. 2585, https://doi.org/10.1007/978-1-0716-2760-0_12,

Table 1
Diagnostic methods for WNV infection in humans

Method	Detectible biological sample	Assay time	Specimen type
Serological testing	WNV-specific antibodies	1–3 days	CSF or serum
Nucleic acid amplification test, such as PCR	WNV RNA	3–4 h	Urine, whole blood, serum, tissue
PRNT	WNV-neutralizing antibodies	3–5 days	Typically plasma or serum
Virus isolation	Infectious whole virus	3–10 days	Any positive patient sample
Immunohistochemistry	WNV antigen	1–3 days	Tissue

WNV-induced signs and symptoms with other endemic arboviruses like Zika virus. Thus, laboratory detection methods such as ELISA, plaque reduction and neutralization test (PRNT), and nucleic acid amplification tests (NAATs) are critical to help differentiate WNV infection from other pathogens (Table 1). This chapter will discuss clinical diagnosis of WNV in humans, common diagnostic tools used, and upcoming testing measures for WNV detection, focusing on PRNT methodology, as it is critical for confirmatory testing.

1.1 Clinical Diagnosis of WNV in Humans

WNV has a fairly short-term viremic load in serum samples [10–12]; RNA can be detected for up to 15 days after infection. In contract, WNV RNA has been detected for 25 days in urine and greater than 30 days in whole blood after initial onset of symptoms [10]. While not commonly used in clinics, current research suggests that whole blood and urine can extend the time to detection of WNV.

The three most common modes of detection are NAATs, serologic assays, and PRNTs (Table 1). Patients who develop symptoms from WNV infection are most commonly diagnosed by serologic assay detection of anti-WNV antibodies [13]. Anti-WNV IgM antibody levels can be detected within 8 to 21 days after clinical symptom onset [14], making them the best option for detecting acute infection, but these antibodies can persist for over a year in some cases [15, 16]. Anti-WNV IgG testing is important for population-based studies of prevalence and to evaluate seroconversion of patients over time, but not for defining incidence of disease. WNV-specific serologic assays can be performed on acute or convalescent samples of cerebrospinal fluid (CSF) or serum by ELISA. A limitation of serologic assays is the potential for cross-reactivity with other closely related flaviviruses. To overcome this limitation, patients with positive serologic assays should have confirmatory testing using PRNTs, as the use of virus allows for increased anti-WNV specificity. Alternatively, WNV can be detected using NAATs, which have high sensitivity and specificity, are easy to use, and have a short turnaround time when compared to other assays. While serological assay can detect anti-WNV antibodies in the serum or CSF of patients, NAATs detect WNV RNA in many samples, including CSF, serum, whole blood, or tissues. NAATs are particularly helpful in patients who are immunocompromised and may not generate an adequate host immune response for detection by serologic or PNRT assays [17]. Furthermore, NAATs are a critical tool in public health screening, as it enables early detection of WNV in blood bank and transplant donors [18]. Other testing modalities such as immunohistochemistry (IHC) on formalin-fixed tissue and viral cultures are also permitted by CDC for diagnostics in the United States, but these assays are time intensive and can be costly [19]. As multiple commercially available options exist for ELISA antibody testing and for NAATs, the remainder of this chapter will focus on the PRNT.

2 Materials

1. Biosafety cabinet (*see* **Note 1**).
2. 37°C incubator at 5% CO_2.
3. Disinfectant, such as Cavicide, 70% ethanol, or 10% bleach.
4. Chemical waste container.
5. Vero CCL-81 cells.
6. Sera to be tested.
7. West Nile virus.
8. Dubecco's Modified Eagle Medium (DMEM) with 2% fetal bovine serum (FBS).
9. Complete DMEM medium (10% FBS, 1× penicillin-streptomycin, L-glutamine, 1× nonessential amino acids).
10. 2× Minimum Essential Medium (MEM).
11. 1.5% Agarose, mixed by combining 1.5 g of molecular biology grade agarose for every 100 mL of sterile water.
12. 50-mL conical tubes.
13. 96-Well low-binding plates.
14. 6-Well or 12-well plates.
15. Serological pipettes.
16. Micropipettes and filter tips.
17. 10% Neutral buffered formalin, commercially available.
18. 0.5% crystal violet, commercially available.

3   Methods

PRNT is used as both a confirmatory method and titration of WNV-specific neutralizing antibodies. Commonly performed using sera, collected patient samples are challenged with WNV, and plaque formation is quantitated. While PRNT is the "gold standard," it can be labor intensive and requires days for results to be available. The following is a general protocol of WNV PRNT (*see* **Note 2**).

1. Seed Vero CCL-81 cells onto tissue culture multi-well plates 1 day prior to the assay. Suggested seeding densities are as follows:

   1.1. 6-Well plates: 2E5 cells/well, 2 mL of complete DMEM

   1.2. 12-Well plates: 1E5 cells/well, 1 mL of complete DMEM

2. Heat-inactivate sera at 56°C for 30 min.

3. In a non- or low-binding 96-well plate, add DMEM containing 2% FBS for serial dilution of the heat-inactivated sample or samples (*see* **Note 3**).

4. Transfer heat-inactivated sera into appropriate wells in the first column of 96-well plate to create the desired dilution, and mix well. Repeat until all dilutions are complete.

5. Thaw West Nile virus stock of known titer, and mix the stock by gentle vortexing in a biosafety cabinet. Dilute virus in DMEM with 2% FBS to desired concentration.

   5.1. Generally, a 12-well assay should have 30–50 plaques, while a 6-well assay should have 40–60 plaques.

6. Carefully add the viral dilution in a 1:1 ratio to the diluted serum to each well, mixing gently by pipetting, avoiding splashes and bubbling.

7. Cover the 96-well plate now containing virus and serum in a 37 °C incubator for 1 h.

8. Remove cell culture media from cells seeded ~~on your chosen~~ multi-well plate, and add 100 μl of the serum-virus mixture from the 96-well plate to each 12-well (200 μL if using a 6-well plate), starting with the most-dilute sample.

9. Apply inoculum carefully near the plate wall to prevent splashing. Gently tilt the plate to evenly distribute inoculum over cells.

10. Place the plates in the 37°C incubator for 1 h. Gently tilt the culture plate every ~15 min to prevent drying and encourage even spread.

11. Prepare overlay media: one part each of the complete DMEM, 2×MEM, and 1.5% agarose in water. Complete DMEM and

$2 \times$ MEM should be warmed to approximately 37°C using a water or bead bath before the 1.5% agarose solution is added. 1.5% agarose can be added and kept at $42–45^\circ$C for up to 30 min before adding to the cells.

12. Using a serological pipet, carefully add warm (~37°C) overlay mixture to each well of the infected multi-well plates (1.5 mL for 12-well plates, 3 mL for 6-well plates), avoiding splashes. After the overlay has been applied, leave the plates in the biosafety cabinet for a few minutes to allow the overlay mixture to solidify.

13. Incubate the cells at 37°C in the incubator with 5% CO_2.

14. Clean up following the appropriate biosafety guidelines for your laboratory and institution.

15. After plaques develop, typically 3 days after incubation, fix cells by adding 10% neutral buffered formalin to each well. Leave to fix for at least 2 h at room temperature.

16. Collect formalin and formalin-soaked agarose overlay plugs into a chemical waste container. Add enough 0.5% crystal violet to cover the cells in each well, and let sit for 3–5 min.

17. Remove crystal violet stain, and dispose of appropriately per your laboratory chemical safety plan. Rinse wells until no crystal violet remains and leave plates to dry.

18. Once plates are dry, count plaques.

19. Calculate the percent neutralization by dividing the sample well counts by the control counts for each dilution. Subtract this number from 1 and multiply by 100 to get the percent (*see* **Note 4**).

20. Endpoint titers are assigned as the greatest dilution in which $\geq$90% (PRNT90) or $>$ 50% (PRNT50) neutralization of the challenge virus was achieved.

4 Notes

1. Aseptic technique should be used when handling all materials to prevent contamination. It is suggested that only cells that have tested negative for mycoplasma are used for the assay.

2. All procedures should be carried out in a biosafety cabinet until the fixation step is complete. It is suggested that all samples be tested in duplicate, with the virus-only controls performed in at least triplicate wells.

3. Dilutions should be a minimum of twofold serial dilutions, while threefold and fourfold serial dilutions may be preferred. For example, if you would like to test twofold serial dilutions,

beginning with 1:10, at least 50 µL of serum should be heat inactivated. In the first dilution, you will add 120 µL of DMEM containing 2% FBS, and all other wells will contain 75 µL of DMEM containing 2% FBS. To the first dilution, you will add 30 µL of heat-inactivated sera and mix well. Then, 75 µL of sera:media mixture will be added to the next well. At the last dilution, 75 µL will be discarded. The tips should be discarded after each dilution.

4. The formula to calculate the percent neutralization is (1 − average plaque counts of samples incubated with virus/average plaque counts of virus-only control) * 100. If you have a control well with 50 plaques and a sample well with 5 plaques, (1–5/50) * 100 = 90%.

References

1. Ronca SE, Murray KO, Nolan MS (2019) Cumulative incidence of West Nile virus infection, continental United States, 1999–2016. Emerg Infect Dis 25(2):325–327. https://doi.org/10.3201/eid2502.180765

2. Mostashari F, Bunning ML, Kitsutani PT, Singer DA, Nash D, Cooper MJ, Katz N, Liljebjelke KA, Biggerstaff BJ, Fine AD, Layton MC, Mullin SM, Johnson AJ, Martin DA, Hayes EB, Campbell GL (2001) Epidemic West Nile encephalitis, New York, 1999: results of a household-based seroepidemiological survey. Lancet 358(9278):261–264. https://doi.org/10.1016/S0140-6736(01)05480-0

3. Bode AV, Sejvar JJ, Pape WJ, Campbell GL, Marfin AA (2006) West Nile virus disease: a descriptive study of 228 patients hospitalized in a 4-county region of Colorado in 2003. Clin Infect Dis 42(9):1234–1240. https://doi.org/10.1086/503038

4. Cook RL, Xu X, Yablonsky EJ, Sakata N, Tripp JH, Hess R, Piazza P, Rinaldo CR (2010) Demographic and clinical factors associated with persistent symptoms after West Nile virus infection. Am J Trop Med Hyg 83(5):1133–1136. https://doi.org/10.4269/ajtmh.2010.09-0717

5. Hasbun R, Garcia MN, Kellaway J, Baker L, Salazar L, Woods SP, Murray KO (2016) West Nile virus retinopathy and associations with long term neurological and neurocognitive sequelae. PLoS One 11(3):e0148898. https://doi.org/10.1371/journal.pone.0148898

6. Murray KO, Garcia MN, Rahbar MH, Martinez D, Khuwaja SA, Arafat RR, Rossmann S (2014) Survival analysis, long-term outcomes, and percentage of recovery up to 8 years post-infection among the Houston West Nile virus cohort. PLoS One 9(7):e102953. https://doi.org/10.1371/journal.pone.0102953

7. Murray KO, Nolan MS, Ronca SE, Datta S, Govindarajan K, Narayana PA, Salazar L, Woods SP, Hasbun R (2018) The neurocognitive and MRI outcomes of West Nile virus infection: preliminary analysis using an external control group. Front Neurol 9:111. https://doi.org/10.3389/fneur.2018.00111

8. Murray KO, Resnick M, Miller V (2007) Depression after infection with West Nile virus. Emerg Infect Dis 13(3):479–481. https://doi.org/10.3201/eid1303.060602

9. Weatherhead JE, Miller VE, Garcia MN, Hasbun R, Salazar L, Dimachkie MM, Murray KO (2015) Long-term neurological outcomes in West Nile virus-infected patients: an observational study. Am J Trop Med Hyg 92(5):1006–1012. https://doi.org/10.4269/ajtmh.14-0616

10. Gorchakov R, Gulas-Wroblewski BE, Ronca SE, Ruff JC, Nolan MS, Berry R, Alvarado RE, Gunter SM, Murray KO (2019) Optimizing PCR detection of West Nile virus from body fluid specimens to delineate natural history in an infected human cohort. Int J Mol Sci 20(8). https://doi.org/10.3390/ijms20081934

11. Lustig Y, Mannasse B, Koren R, Katz-Likvornik S, Hindiyeh M, Mandelboim M, Dovrat S, Sofer D, Mendelson E (2016) Superiority of West Nile virus RNA detection in whole blood for diagnosis of acute infection. J Clin Microbiol 54(9):2294–2297. https://doi.org/10.1128/JCM.01283-16

12. Tilley PA, Fox JD, Jayaraman GC, Preiksaitis JK (2006) Nucleic acid testing for West Nile virus RNA in plasma enhances rapid diagnosis of acute infection in symptomatic patients. J Infect Dis 193(10):1361–1364. https://doi.org/10.1086/503577

13. Martin DA, Biggerstaff BJ, Allen B, Johnson AJ, Lanciotti RS, Roehrig JT (2002) Use of immunoglobulin m cross-reactions in differential diagnosis of human flaviviral encephalitis infections in the United States. Clin Diagn Lab Immunol 9(3):544–549. https://doi.org/10.1128/cdli.9.3.544-549.2002

14. Rawlins ML, Swenson EM, Hill HR, Litwin CM (2007) Evaluation of an enzyme immunoassay for detection of immunoglobulin M antibodies to West Nile virus and the importance of background subtraction in detecting nonspecific reactivity. Clin Vaccine Immunol 14(6):665–668. https://doi.org/10.1128/CVI.00480-06

15. Papa A, Anastasiadou A, Delianidou M (2015) West Nile virus IgM and IgG antibodies three years post- infection. Hippokratia 19(1):34–36

16. Murray KO, Garcia MN, Yan C, Gorchakov R (2013) Persistence of detectable immunoglobulin M antibodies up to 8 years after infection with West Nile virus. Am J Trop Med Hyg 89(5):996–1000. https://doi.org/10.4269/ajtmh.13-0232

17. Lee BY, Biggerstaff BJ (2006) Screening the United States blood supply for West Nile virus: a question of blood, dollars, and sense. PLoS Med 3(2):e99. https://doi.org/10.1371/journal.pmed.0030099

18. Zou S, Foster GA, Dodd RY, Petersen LR, Stramer SL (2010) West Nile fever characteristics among viremic persons identified through blood donor screening. J Infect Dis 202(9):1354–1361. https://doi.org/10.1086/656602

19. Nasci RS, White DJ, Stirling H, Oliver JA, Daniels TJ, Falco RC, Campbell S, Crans WJ, Savage HM, Lanciotti RS, Moore CG, Godsey MS, Gottfried KL, Mitchell CJ (2001) West Nile virus isolates from mosquitoes in New York and New Jersey, 1999. Emerg Infect Dis 7(4):626–630. https://doi.org/10.3201/eid0704.010404

Chapter 13

Detection and Analysis of West Nile Virus Structural Protein Genes in Animal or Bird Samples

Gili Schvartz, Sharon Karniely, Roberto Azar, Areej Kabat, Amir Steinman, and Oran Erster

Abstract

West Nile virus (WNV) is an important zoonotic pathogen, which is detected mainly by identification of its RNA using PCR. Genetic differentiation between WNV lineages is usually performed by complete genome sequencing, which is not available in many research and diagnostic laboratories. In this chapter, we describe a protocol for detection and analysis of WNV samples by sequencing the entire region of their structural genes capsid (C), preM/membrane, and envelope. The primary step is the detection of WNV RNA by quantitative PCR of the NS2A gene or the C gene regions. Next, the entire region containing the structural protein genes is amplified by PCR. The primary PCR product is then amplified again in parallel reactions, and these secondary PCR products are sequenced. Finally, bioinformatic analysis enables detection of mutations and classification of the samples of interest. This protocol is designed to be used by any laboratory equipped for endpoint and quantitative PCR. The sequencing can be performed either in-house or outsourced to a third-party service provider. This protocol may therefore be useful for rapid and affordable classification of WNV samples, obviating the need for complete genome sequencing.

Key words West Nile virus, Quantitative PCR, Nested PCR, Sequencing

1 Introduction

West Nile virus (WNV, family *Flaviviridae*, genus *Flavivirus*) is the etiological agent of West Nile fever (WNF) disease. It is a mosquito-borne virus with an unusually large host range of reptiles, avian species, and mammals [1–3]. WNV is currently abundant in all continents except Antarctica and causes occasional morbidity waves among birds, horses, and humans in infected regions [3, 4]. Diagnosis of WNV infection was mainly performed using serological tests in past years and is still used together with a gradual increase in molecular-based tests [5, 6]. Molecular analysis of the WNV is important for epidemiological studies, as well as to understand its pathogenicity and its response to the host immune system

Fengwei Bai (ed.), *West Nile Virus: Methods and Protocols*,
Methods in Molecular Biology, vol. 2585, https://doi.org/10.1007/978-1-0716-2760-0_13,
© The Author(s), under exclusive license to Springer Science+Business Media, LLC, part of Springer Nature 2023

[7–10]. Rapid examination of suspected samples is currently performed using quantitative reverse transcription PCR (RT-qPCR or qPCR), which can provide definite results for a relatively large number of samples within a few hours from the sample collection. Quantitative PCR does not usually provide detailed information on the classification of the sample, and it cannot be used to detect single-base differences between different samples. Whole genome sequencing is currently the most comprehensive tool for such analysis, providing an unparalleled amount of genetic information [11]. Whole genome sequencing requires resources and qualified dedicated personnel, as well as specific equipment, and these are not currently available in most laboratories worldwide. In this chapter, we describe a protocol for analysis of the WNV genome part that includes the structural genes capsid (C), pre-membrane/membrane (preM/M), and envelope (E). These proteins are exposed to the host immune system when viral particles are released from infected cells. Changes in these proteins are therefore important for understanding host-pathogen interactions related to cell entry and immune response [9, 10]. This protocol can be performed in any molecular biology laboratory, up to the sequencing step. The sequencing reaction can either be performed in-house if the laboratory is sufficiently equipped or can be outsourced, as a standard sequencing service from a commercial provider. This protocol therefore represents an intermediate method that does not require next-generation sequencing facility, but provides valuable information that cannot be obtained using only quantitative or endpoint PCR.

The workflow of the protocol is described in Fig. 1.

2   Materials

2.1   General Laboratory Materials and Equipment

1. Pipettes set suitable for work volumes from 0.5 μL to 1 mL.
2. 1.5-mL microcentrifuge tubes.
3. Thin-wall 0.2-mL endpoint PCR tubes.
4. Thin-wall 0.1-mL real-time PCR tubes or strips.
5. Surgical blades.
6. Heating/cooling block for 1.5-mL tubes.
7. PCR-grade H_2O or PCR-grade TE buffer suitable for nucleic acid resuspension.

The following laboratory equipment is required for performing the protocol:

1. 4 °C refrigerator or cooler container adjusted to 4 °C.
2. −20 °C or −80 °C freezer.

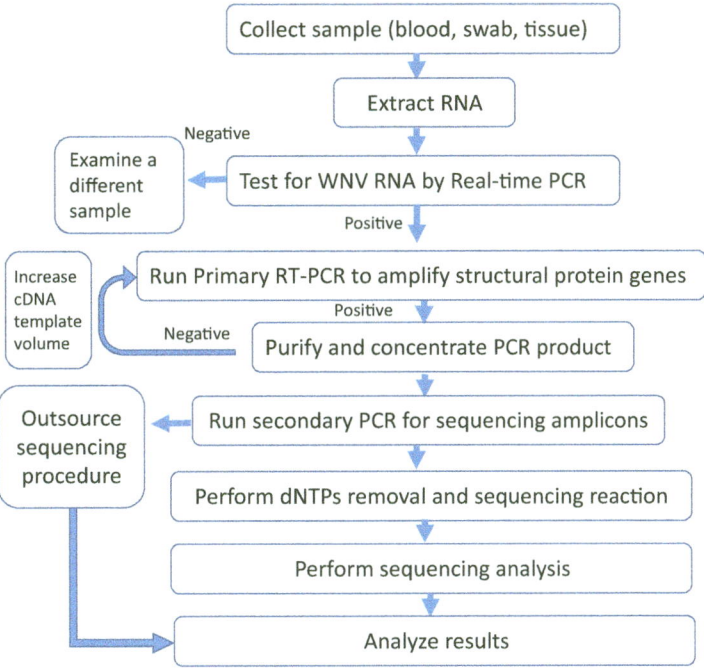

Fig. 1 General scheme of the workflow for analyzing the structural protein genes of West Nile virus

3. Spectrophotometer or fluorimeter compatible with nucleic acid quantification.

4. Real-time PCR cycler.

5. Endpoint PCR cycler.

6. Agarose gel electrophoresis apparatus.

7. Optical imager for agarose gel electrophoresis analysis.

8. Cooled centrifuge for 1.5-mL or 2-mL tubes.

9. Racks for collection tubes, 1.5-mL tubes, 0.2-mL tubes, and 0.1-mL tubes.

10. Microwave oven or a hot plate for agarose gel preparation.

11. For sequence analysis – PC or Mac.

2.2 Sample Collection and Preparation

1. Collection dry swabs and tubes or collection tubes containing transfer medium such as viral transport medium (VTM) or universal transport medium (UTM).

2. Sterile mortar and pestle.

3. Phosphate buffer saline (PBS, molecular biology-grade).

2.3 RNA Extraction

1. Total nucleic acid or RNA extraction kit.

2. RNAse-free barrier tips.

3. RNAse-free tubes.

2.4 Real-Time PCR

1. Master mix: Real-time PCR should be performed using a one-step probe real-time PCR protocol. For considerations and recommendations, *see* **Note 1**.

2. Real-time PCR primers and probes: Detect WNV RNA in the sample using two separate reactions, with the following primers and probe sets. The first reaction was described by Eiden et al. [5] and targets the NS2A gene. Use the following primers and probes: forward primer: 5'-GGGCCTTCTGGTCGTGTTC-3' and reverse primer: 5'-GATCTTGGCYGTCCACCTC-3'; and probe: 5'-6-FAM-CCACCCAGGAGGTCCTTCGCAA-BHQ1–3'. The second reaction was described by Schvartz et al. [12], targets the capsid (C) protein gene, and includes the following components: forward primer: 5'-GTGCTGG ATCGATGGAGAGG-3' and reverse primer: 5'-GTGCTGGA TCGATGGAGAGG-3' and probe: 5'-6-FAM-CAAACAGCG ATGAAACACCTTC TG-BHQ1-3'.

    Perform the amplification reaction in any certified real-time PCR thermal cycler.

2.5 Endpoint PCR

2.5.1 Primary Endpoint PCR Setup

1. For cloning of the entire region that encodes the structural genes capsid, preM/M, and envelope, use the following primers: 30 forward: 5'- CAACAATTAACACAGTGC-GAGCTG-3' and 2494 reverse 5'-CATTGTGTATGAACACTCCACTTC-3'.

2. Use a PCR master mix compatible with amplification of long (≥3 Kb) products.

3. Any certified thermal cycler can be used for the amplification reaction.

2.6 Agarose Gel Electrophoresis

1. Tris-acetic acid (TAE)-EDTA buffer preparation: Prepare TAE buffer 50× stock solution, as follows: Dissolve 242 g Tris base in 57.1 mL of glacial acetic acid and 200 mL of 0.1 M EDTA pH 7.4, and add H_2O to a final volume of 1 L. Dilute the desired volume of the stock TAE buffer before use.

    Alternatively, TAE buffer and pre-casted gels can be purchased from many manufacturers.

2. Molecular biology-grade agarose for gel electrophoresis.

3. DNA loading dye for gel electrophoresis.

4. Molecular DNA size marker (*see* **Note 2**).

2.7 Purification and Concentration of PCR Products

1. Concentrate and purify the resulting PCR product using a PCR/gel purification kit, according to the manufacturer's instructions. For recommendations, please *see* **Note 3**.

2.8 Secondary PCR for Sequencing

Use the primary ~2.5 Kb product as a template to amplify the sequencing segments, using the primer pairs that are detailed in Table 1.

Table 1
Primers used for sequencing amplicons PCR

Name	Sequence 5'- > 3'	Reaction	Comments
WNV 30 Fwd	CAACAATTAACACAGTGCGAGCTG	Amplicon 1	
WNV 958 rev	TCCAGGAAGTCTCTGTTRCTCATTC	Amplicon 1	
WNV 805 Fwd	TGGTGAARACAGAATCATGGAT	Amplicon 2	
WNV 1435 T Rev	GGAGT**GAT**G**CC**GAATCT**C**CC**T**G	Amplicon 2	Variable bases are in bold
WNV 1435A Rev	GGAGT**TAT**A**CT**GAATCT**T**CC**A**G	Amplicon 2	Variable bases are in bold
WNV 1269 G Fwd	GG**A**AA**G**GG**G**AGCATTGACACATG	Amplicon 3	Variable bases are in bold
WNV 1269 A Fwd	GG**C**AA**A**GG**A**AGCATTGACACATG	Amplicon 3	Variable bases are in bold
WNV 2107 T Rev	TT**T**ATCTG**C**TGTTCTCCTCTT	Amplicon 3	Variable bases are in bold
WNV 2107G Rev	TT**G**ATCTG**T**TGTTCTCCTCTG	Amplicon 3	Variable bases are in bold
WNV 1769 Fwd	TCAAGCAACACTGTSAAGTTGAC	Amplicon 4	
WNV 2494 Rev	CATTGTGTATGAACACTCCACTTC	Amplicon 4	

Variable bases are highlighted in bold

2.9 Preparation of PCR Products for Sequencing

1. If the sequencing is performed in-house, purify the products of the secondary PCR using either a PCR purification kit, as described in Subheading 2.7, or an alternative product that removes unbound dNTPs and is compatible with downstream sequencing.

2. If the sequencing is performed by a third party, such as a commercial-sequencing service provider or a collaborator laboratory, use the PCR/gel purification kit described in Subheading 2.7, to purify and then quantify the PCR product.

2.10 Sequence Analysis

Free sequence analysis tools are available online, to analyze chromatogram sequence files. The following links enable downloading of some of them:

Chromaseq: http://mesquiteproject.org/packages/chromaseq/manual/.

Chromas: http://technelysium.com.au/wp/chromas/.

4peaks: https://nucleobytes.com/4peaks/ (for Mac only).

Sequence scanner: http://resource.thermofisher.com/page/WE28396_2/.

Ugene: http://ugene.net/.

Snapgene: https://www.snapgene.com/snapgene-viewer/.

Alternatively, licensed tools can be used, such as Sequencher (https://www.genecodes.com/). Geneious (https://www.geneious.com/).

3   Methods

3.1   Sample Collection

Collect the sample material from infected animals and birds using dry or preservation medium-containing collection tubes. Collect blood samples into EDTA-supplemented tubes to avoid blood clotting. For comments and suggestions regarding collection media, please *see* **Note 4**. If the sample material is tissue, proceed to Subheading 3.2. If the sample material is blood, proceed to Subheading 3.3 and follow the extraction kit manufacturer's instructions.

3.2   Tissue Sample Preparation

1. Remove tissue from suspected animals and birds using sterile surgical blade. Slice the tissue sample for improved homogenization. Once the tissue is excised, keep it cooled and store it at −80 °C if not processed immediately.

2. Tissue sample preparation for RNA extraction: Homogenize the sample by grinding it using a cooled (4 °C) sterile mortar and pestle, or using a mechanical homogenizer. Add cold (4 °C) phosphate buffer saline in a mass/volume ratio of 1:7. For example, homogenize a tissue weighing 1 g in 7 mL of PBS. Homogenize the sample and incubated at 4 °C, or on ice, for 15 min.

3. Centrifuge the homogenate for 10 min at 1100 g at 4 °C and collect the liquid fraction for RNA extraction.

4. Swab sample preparation: If the swab was incubated in a transfer medium, use the resuspension directly for RNA extraction. If a dry swab was used, soak the swab in 0.3–0.5 PBS for 10 min and then proceed to RNA extraction.

5. Blood sample preparation: Dilute the blood 1:5 in cold PBS and use directly for RNA extraction.

3.3   RNA Extraction

Any nucleic acid extraction kit can be used, providing that it is intended to extract RNA. The instructions herein are general and should be modified depending on the specific product that is used, according to the manufacturer's instructions.

1. Add 300 μL of buffer lysis and 7 μL of Carrier RNA solution into a 1.5-mL tube.

2. Transfer 100 μL of the sample into the tube containing lysis buffer. If the initial sample volume is less than 100 μL, add PBS to a final volume of 100 μL.

3. Vortex the mixture for 10 s.

4. Incubate the mixture for 10 min at room temperature.

5. Add 350 μL of the kit's binding buffer to the mixture and vortex for 10 s. If the sample volume is larger, increase the

binding buffer volume accordingly to keep a ratio of 1:1 with the sample mixture. Do not centrifuge the sample at this stage to avoid nucleic acid precipitation.

6. Transfer 750 μL of the mixture to the kit spin column.

7. Centrifuge at ≥10,000 g for 30 s at room temperature. Discard the filtered liquid, and place the spin column back into the same tube. If the sample mixture volume is larger than 750 μL, repeat this step until all the mixture has passed through the column.

8. Add 500 μL of the kit wash buffer to the spin column and centrifuge at ≥10,000 g for 30 s at room temperature. IMPORTANT: In most kits, you need to make sure that ethanol was added to the wash buffer bottle before adding it to the column. Discard the filtered liquid and place the spin column back into the same tube. Repeat this step again to complete two washing rounds.

9. Centrifuge at full speed for an additional 1 min at room temperature to remove residual wash buffer. Examine the filter column after centrifugation to ensure that no residual buffer is left in it. The column must be dry to avoid inhibition of downstream applications due to residues contamination.

10. Transfer the spin column to a labeled, RNase-free 1.5-mL tube, and add 20–50 μL of nuclease-free water or kit elution buffer, if provided, to the center of the spin column membrane. Incubate at room temperature for 1–5 min.

11. Centrifuge the spin column for 1 min at ≥10,000 g and discard the column.

It is recommended, but not necessary, to measure the RNA concentration to evaluate the extraction efficiency before performing the PCR analysis.

3.4 Real-Time PCR

In order to identify the presence of WNV RNA in the examined sample, run a real-time PCR test (quantitative PCR or qPCR) that can detect WNV RNA, regardless of its lineage. We use two different reactions, as detailed below. For further suggestions and tips for successful real-time PCR assays, please *see* **Notes 1** and **5**.

1. For the NS2A test, use the NS2A primers that are described in Subheading 2.4, **item 2**. For the C gene test, use the capsid gene primers described therein. Both assays are assembled using the same concentrations and are run using the same thermal protocol.

Assemble the reaction mix by combining the following components per one reaction:

2× reaction buffer: 12.5 μL, RT enzyme: 1 μL, forward and reverse primers (10 μM stock): 1 μL each (final

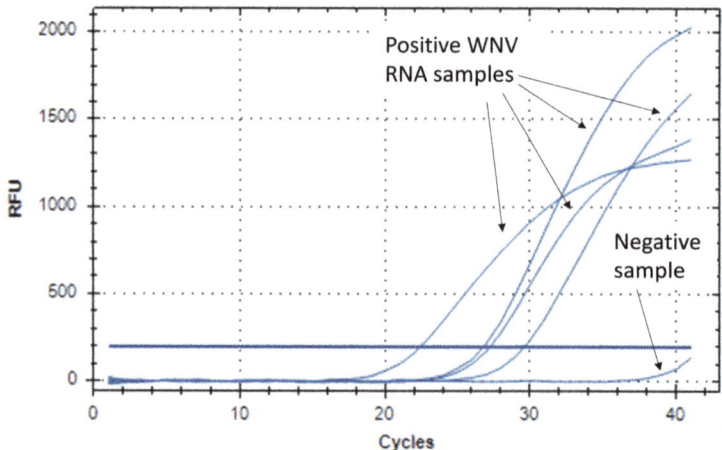

Fig. 2 Real-time PCR analysis of WNV samples. Positive samples generate a clear signal with fluorescence intensity above 500 RFU (relative fluorescence units), while a negative signal results in complete absence, or a very weak signal, below 200 RFU

concentration of 400 nM), probe (10 μM stock): 0.4 μL (final concentration of 160 nM), RNA: 5 μL, and PCR-grade H$_2$O 5.1 μl to a final volume of 25 μL.

2. Set the optical thermal cycler to run the reaction using the following conditions: 45 °C for 10 min (RT synthesis), 95 °C for 10 min (DNA polymerase activation), 40 cycles of 45 °C for 15 s (denaturation), and 60 °C for 45 s (annealing and extension). Set the instrument fluorescence read at the 60 °C step.

3. Analyze the results using the instrument analysis software and identify positive samples. See the expected results in Fig. 2 in Subheading 3.10. If the sample is positive for WNV RNA, proceed to Subheading 3.4, for amplification of the structural genes.

3.5 Primary Endpoint PCR

The RT-PCR protocol below is a generic protocol. For considerations and recommendations on performing the RT-PCR, please *see* **Note 5**.

1. Prepare the following reaction mix to amplify the entire WNV structural genes region: 2× reaction master mix, 20 μL; RT enzyme, 1.5 μL; 30 forward and 2494 reverse primers (10 μM stock), 1.5 μL each; RNA, 4 μL; and PCR-grade H$_2$O, 11.5 μL to 40 μL final volume. The details of the primers are described in Subheading 2.6, **item 1**.

2. Set the thermal cycler to the following program: 45 °C for 15 min (RT synthesis); 95 °C for 2 min (RT inactivation, DNA polymerase activation); and 40 amplification repeats:

95 °C for 10 s, 59 °C for 10 s, and 72 °C for 60 seconds; 1 repeat: 72 °C for 5 min.

Upon completion of the amplification reaction, proceed to PCR analysis or store the PCR tubes in −20 °C.

3.6 Analysis of PCR Products Using Agarose Gel Electrophoresis

Examine the endpoint PCR products using agarose gel electrophoresis. Agarose gel can be purchased, or prepared in-house. The following section describes in-house preparation. For considerations and recommendations, *see* **Note 6**.

1. To prepare the agarose gel, add 1 g of molecular biology-grade agarose per 100 mL of 1× TAE in a capped glass bottle, and gradually warm in a microwave oven or on a hot plate until the agarose is completely dissolved.

2. Mix carefully to dissolve any remaining clumps, and add DNA staining dye, one drop per 50-mL gel.

3. Cast the gel in an agarose gel electrophoresis apparatus, and insert the corresponding lane combs, according to the gel apparatus manufacturer's instructions.

4. Let the gel solidify at room temperature.

5. Once the gel is ready to use, carefully remove the gel combs, and fill the gel apparatus with 1× TAE buffer.

6. If you are using a PCR mix that contains a loading dye (pre-stained mix), load 6 μL of the reaction directly onto the gel. If the PCR mix does not contain a loading dye, mix 6 μL of the PCR product with 5 μL loading dye, and then load the samples onto the gel.

7. Load 5 μL of the DNA size marker ("DNA ladder"), and run the samples at 120 V for 20 min.

8. Examine the gel in an optical imager using UV light illumination. The desired product should be resolved in parallel with the 2.5 Kb band of the size marker (*see* Fig. 3). If the size marker is not properly separated, continue running the gel for another 10–15 min, and examine it again. For recommendations and suggestions on gel preparation and electrophoresis analysis, please *see* **Note 6**.

3.7 PCR Product Purification

The resulting DNA product can be purified using any commercial kit for PCR product purification. The procedure below is a generic protocol that should be modified as per the specific product that is used. For considerations and recommendations on PCR product purification, *see* **Note 7**.

1. If necessary, add PCR-grade H_2O to the PCR product, to adjust the volume to 50 μL.

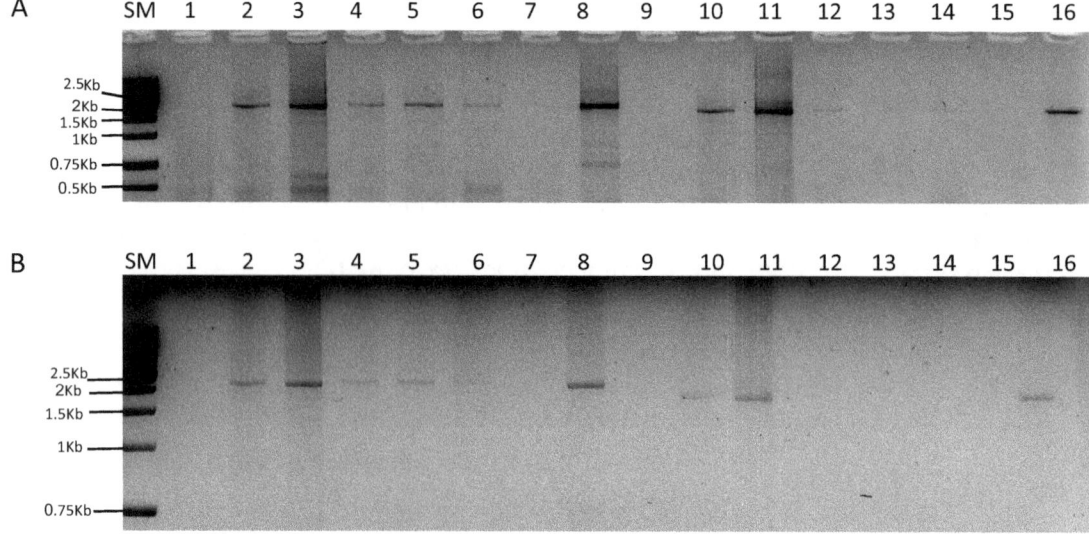

Fig. 3 Primary WNV structural protein gene amplification. Agarose gel electrophoresis of 16 WNV samples following amplification with primers 30 forward and 2494 reverse. (**a**) Separation after 30 min. (**b**) Separation after 40 min. SM, size marker

2. Mix the PCR product with binding buffer in a ratio of 1:2. For example, add 100 μL of binding buffer to 50 μL of PCR product.

3. Load the mix on the spin column and centrifuge at 11,000 g for 30 s. Load up to 700 μL. if the mix volume is larger, re-load the column after the first centrifugation, and then repeat this step, until all the PCR-binding buffer mixture is passed through the column.

4. Add 500 μL of wash buffer to the column and centrifuge at 11,000 g for 30 s. Discard the flow-through and repeat this step. IMPORTANT: In most kits, you need to make sure that ethanol was added to the wash buffer bottle before adding it to the column.

5. Dry the column from residual wash buffer by centrifugation at 11,000 g for 1–2 min (at least 1 min).

6. Carefully transfer the column to a labeled, RNAse-free 1.5 tube, add 30 μL elution buffer, and incubate at room temperature (15–25 °C) for 5 min.

7. Centrifuge at 11,000 g for 1 min and measure the concentration of the eluted DNA (*see* **Note 7**).

3.8 Secondary PCR for Sequencing

To facilitate sequencing of the entire structural genes coding region by Sanger method, the product of the primary PCR is further amplified in four segments, each between 600 and 1000 bp. In

Table 2
Thermal profile of the secondary PCR amplification

Component	Volume (μL)
Reaction buffer ×2	30
Forward primer	2
Rev primer	2
H_2O	24
Primary PCR product	4
Total reaction volume	60

the secondary PCR, the primary PCR product is used as template to allow sufficient amplification for downstream sequencing of each segment.

1. Use the following combinations of primers, to generate each segment:

   Amplicon 1: 30 Fwd + 958 Rev. Product size: 928 bp.

   Amplicon 2: 805 Fwd + 1435TRev + 1435ARev. Product size: 630 bp.

   Amplicon 3: 1269A Fwd + 1269G Fwd + WNV 2107 T Rev.+ WNV 2107G Rev. Product size: 838 bp.

   Amplicon 4: 1769 Fwd + 294 9Rev. Product size: 725 bp.

   Prepare four mixes, for each sequencing segment; each contains a different set of primers. The mix components are detailed in Table 2. For considerations and recommendations on performing the secondary PCR, *see* **Note 5**.

2. Assemble the reaction components for each reaction as detailed in the table above, and divide each reaction into two 0.2-μL-thin-wall PCR tubes.

3. Place the tubes in the PCR thermal cycler, and set the reaction parameters as detailed in Table 3.

4. Analyze the PCR results using gel electrophoresis as detailed in Subheading 3.6, and document the results using an optical imager.

3.9 Sample Preparation for Sequencing

1. If the sequencing reaction is performed by a third party, such as a commercial service provider, perform purification as described in Subheading 3.7. If the sequencing is performed in-house, proceed to Subheading 3.8, **step 4**.

2. Determine the purified product concentration using a spectrophotometer or a fluorimeter compatible for DNA quantification.

Table 3
Thermal profile of the primary PCR amplification

Step	Conditions	Repeats
Initial denaturation	95 °C for 40 s	1
Denaturation	95 °C for 10 s	36
Annealing	60 °C for 15 s	36
Extension	72 °C for 20 s	36
Product end completion	72 °C for 5 min	1
Incubation	8 °C hold	1

3. Prepare the secondary PCR product as instructed by the sequencing provider, and transfer to the sequencing provider. A separate reaction is required for each sequence with each primer. For comments on sequencing preparation, *see* **Notes 8** and **9**.

4. If the sequencing is performed in-house, remove residual components that may interfere with downstream sequencing steps, by using an appropriate commercial kit.

5. Perform the labelling reaction according to the sequencing system manufacturer's instructions. For sequencing system suggestion, *see* **Note 10**.

3.10 Sequence Analysis

In order to analyze the sequencing results, align the resulting sequences with a reference sequence that was annotated and approved by a recognized organization such as NCBI GenBank (www.ncbi.nlm.nih.gov), EMBL nucleotide database (https://www.ebi.ac.uk), or DNA databank of Japan (www.ddbj.nig.ac.jp).

Analysis can be performed using the FASTA sequences (end with .fasta extension), or with the histogram files that are obtained from the sequencing reaction (end with .ab1 extension). Several free online tools are available for sequence analysis, as detailed in Subheading 2.10. In order to identify sequence variations that lead to amino acid changes in the protein sequence, the nucleotide sequence needs to be translated to amino acids (*see* **Note 11**).

3.11 Expected Results

1. Real-time PCR analysis: Positive WNV sample should generate a clear signal in the PCR, in contrast to a negative sample, with no signal. Such examples are shown in Fig. 2.

2. Primary PCR: The primary WNV structural genes PCR should result in amplification of a ~2.5 Kb product and may or may not be visible by agarose gel electrophoresis. As shown in Fig. 3, some samples contained enough RNA to generate a clear band, and some may not. The faint bands in lanes 12–14 can be

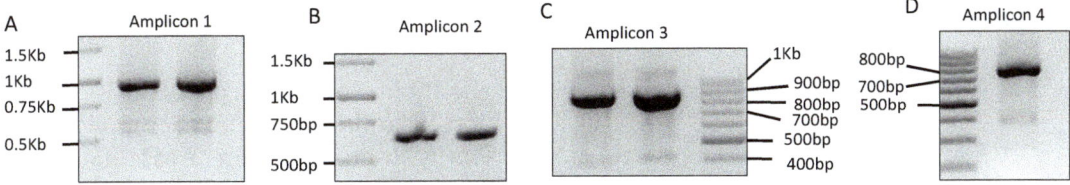

Fig. 4 Secondary amplification of WNV structural protein genes. (**a**) Amplification of segment 1 (930 bp). (**b**) Amplification of segment 2 (630 bp). (**a**) Amplification of segment 3 (840 bp). (**b**) Amplification of segment 4 (725 bp). The size markers are indicated for each segment

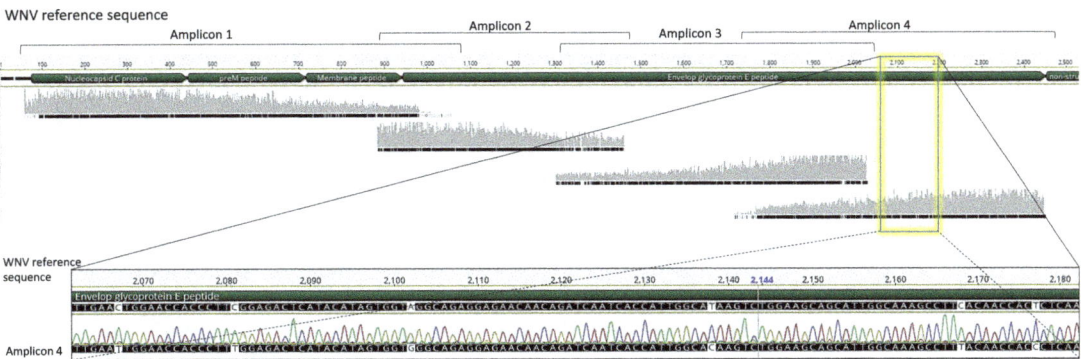

Fig. 5 Analysis of secondary PCR products sequencing. Combined analysis of the four PCR products (amplicons) sequences, mapped to a reference WNV sequence. The enlarged inset shows the histograms ("peaks") of a part of the amplicon 4 sequence aligned with the reference sequence envelope gene

detected when imaging is performed after 30 min or less (Fig. 3a) but are no longer detectable after the additional 10 min separation (Fig. 3b).

3. Secondary PCR: The four reactions designed for sequencing result in products ranging from 600 bp to 950 bp, as detailed in Subheading 3.7, **step 1**. Successful amplification of the four products is shown in Fig. 4. The primary PCR products from two WNV samples were used to amplify the four products.

4. Sequence analysis: Assembly of all four PCR products (amplicon) and their alignment to an annotated reference sequence generate the complete sequencing of the surface protein gene region. The combined alignment of the four amplicons with a reference sequence is shown in Fig. 5. The enlarged inset shows the histogram of a part of amplicon 4 (part of the envelope gene), where some differences between the sequenced sample and the reference sequence were identified. The quality of the sequencing (nucleotide peaks) indicates that the differences are real and do not result from sequencing errors.

4   Notes

1. We successfully used the Bioline SensiFast probe lo-ROX mix (cat. No. BIO-76001, www.bioline.com) and the Applied Biosystems AgPath-ID RT-PCR mix (cat. no. AM1005, www.thermofisher.com). Nevertheless, any commercial real-time PCR master mix may be used, with appropriate adjustments, as instructed by the manufacturer. Most commercial master mixes should be suitable for WNV detection using the protocol described herein. However, the specific properties of each mix should be considered when designing the amplification protocol. The reverse transcription (RT) stage temperature and duration is specific to the RT of the mix and is different in different products. The RT inactivation and DNA polymerase activation may also depend on the specific structure and properties of the enzymes in the kit. Therefore, while the amplification steps should be similar when using different kits, the RT stage and the DNA polymerase activation may vary.

2. In order to ensure the correct size of the PCR products in agarose gel electrophoresis, it is imperative to use a size marker together with the amplified samples. The size marker should be suitable for running samples between 500 bp and 3Kb. Such markers can be purchased from manufacturers such as https://international.neb.com, www.takarabio.com, and www.thermofisher.com/.

3. We use the Machery-Nagel Nucleospin Kit (cat. No. 740609.50, www.mn-net.com), but any PCR purification kit should work. For the preparation of the kit, use molecular biology-grade 100% ethanol, according to the manufacturer's instructions. Alternatively, extract the viral RNA using automated system, such as the PSS MagLEAD (www.pss.co.jp) or NUCLISENS easyMAG (www.biomerieux-usa.com) systems.

4. We used dry swabs or BD transport medium tubes and blood collection tubes (www.bd.com). However, any veterinary-qualified viral or universal transport medium (VTM or UTM) and blood collection tube may be used, as long as blood clotting is prevented.

5. When performing either endpoint or quantitative PCR, the primers and probe concentrations, as well as the reaction conditions, may vary depending on the master mix and the instrument used. The detailed protocol described herein should therefore be used as a guideline, and further optimization may be necessary to obtain optimal results. We recommend using the mix according to the manufacturer's instructions, with the primers and probe concentrations indicated here, as

a starting point. The RT synthesis and the DNA polymerase activation may also depend on the master mix used, and the manufacturer's instructions should serve as a starting point for reaction optimization.

When running a test, we recommend including at least one positive control and one negative control. The positive control should contain WNV RNA and should serve as an indicator for the reaction integrity. The negative control should ideally contain an extraction of a WNV-negative sample. PCR-grade H_2O or TE buffer can also be used. This control should confirm the absence of contamination that can result in a false-positive signal.

Important note for endpoint PCR analysis: If the viral RNA is in low abundance, the PCR may work, but you will not see a product in the electrophoresis analysis. In this case, do not attempt to purify the product. Instead, take 4 μL from the RT-PCR reaction tube, and use it directly as DNA template for the secondary PCR tube.

6. When preparing agarose gel, dissolve the agarose into the buffer by warming the mix bottle with a hot plate or by microwave heating. A hot plate will require less intervention but takes longer, usually 10–20 min, depending on the gel volume. Microwave heating is faster but requires constant attention, since the solution tends to bubble and can cause serious injury in the case of a sudden heat burst. When dissolving agar in a microwave oven, use 30-second intervals, and mix the solution carefully after each interval to facilitate even warming of the buffer. This way the entire volume will be dissolved without heat bursts that can be harmful and lead to serious burns. After the agar is dissolve, add the DNA staining dye. Consider the safety instructions on using the staining dye, as the manufacturer indicates. When casting the gel in the mold, faster cooling is achieved when leaving the apparatus in a chemical hood. Do not let the gel dry for more than 1 h. Overdried gel will crack and cannot be used further.

7. The primary PCR product can be used directly for downstream amplification without the concentration and purification step. However, when amplifying a 2.5-Kb-length region, it is recommended to purify and concentrate the product, for several reasons: (1) The reaction yield is often low, compared with PCR yield of shorter (<2.5 Kb) products, and without concentration, the secondary PCR may be less efficient. (2) When attempting to amplify a 2.5 Kb, shorter products are often generated, and the secondary PCR may be less specific, leading to the generation of multiple nonspecific products, which, in turn, will render the sequencing inefficient. In that case,

consider resolving the primary PCR product on agarose gel, and purify the desired product from the gel. (3) If the primary reaction results in a high product yield, the purification process allows you to measure the product concentration before setting up the secondary reaction. The elution can be performed in 15- to 50-μL elution buffer. Low elution volume will increase the product concentration, but the elution efficiency will be lower. If the product yield is low, elute it in 20 μL, load the eluate AGAIN on the spin column, and repeat the incubation and centrifugation. This should improve the elution efficiency and maintain high product concentration. When imaging the gel-separated samples, record the results before complete separation, as shown in Fig. 3a. This will enable detection of faint bands that may not be visible after extended separation, which provides better resolution, but less sensitivity (Fig. 3b).

8. In order to minimize pipetting errors and maximize accuracy, prepare a master mix with all components, except the primers. Then divide the mix into the required number of tubes, and then add the specific primers for each reaction.

9. Due to the genetic variation among different WNV isolates, some primers will be less efficient for sequencing. Therefore, when preparing the secondary PCR product Amplicon 3 (two 1269 Fwd primers and two 2107 Rev. primers), four sequencing reactions need to be prepared for the same sample, each with a different primer.

10. We used the Applied Biosystems 3500 sequencer system (cat. No. 4406017, www.thermofisher.com), and we perform the labeling reaction with the BigDye terminator kit. However, any Sanger sequencing system should be sufficient to perform the sequencing.

11. The histogram sequencing files provide information on the quality and uniformity of the sequence and are therefore better for analyzing the sequencing results, compared with the FASTA format, which shows only the "letters" of the sequence. However, if a histogram-analyzing software is not available, the sequence can be analyzed using many online tools such as the NCBI BLAST (https://blast.ncbi.nlm.nih.gov) or the EBI Multiple Sequence Alignment tools (https://www.ebi.ac.uk/Tools/msa/). Translation of nucleotide sequences to amino acid sequence can be performed with several online tools, such as https://web.expasy.org/translate, https://www.ebi.ac.uk/Tools/st.

Acknowledgments

The authors wish to acknowledge the members of the divisions of virology and avian diseases in the Kimron Veterinary Institute for their help in cell culture and qPCR experiments and members of Israel Ministry of Health Central Virology Laboratory for their help with the endpoint PCR experiments and sequencing.

References

1. Marra PP, Griffing S, Caffrey C, Kilpatrick MA et al (2004) West Nile virus and wildlife. Bioscience 54(5):393–402. https://doi.org/10.1641/0006-3568(2004)054[0393:WNVAW]2.0.CO;2

2. Klenk K, Snow J, Morgan K, Bowen R et al (2004) Alligators as West Nile virus amplifiers. Emerg Infect Dis 10(12):2150–2155. https://doi.org/10.3201/eid1012.040264

3. Habarugira G, Suen WW, Hobson-Peters J, Hall RA et al (2020) West Nile virus: an update on pathobiology, epidemiology, diagnostics, control and "one health" implications. Pathogens 9(7):589. https://doi.org/10.3390/pathogens9070589

4. Pérez-Ramírez E, Llorente F, Jiménez-Clavero MÁ (2014) Experimental infections of wild birds with West Nile virus. Viruses 6:752–781. https://doi.org/10.3390/v6020752

5. Eiden M, Vina-Rodriguez A, Hoffmann B, Ziegler U et al (2010) Two new real-time quantitative reverse transcription polymerase chain reaction assays with unique target sites for the specific and sensitive detection of lineages 1 and 2 West Nile virus strains. J Vet Diagn Investig 22:748–753

6. Nagy A, Mezei E, Nagy O, Bakonyi T et al (2019) Extraordinary increase in West Nile virus cases and first confirmed human Usutu virus infection in Hungary, 2018. Euro Surveill 24(28):1900038. https://doi.org/10.2807/1560-7917.ES.2019.24.28.1900038

7. Basset J, Burlaud-Gaillard J, Feher M, Roingeard P et al (2020) A molecular determinant of West Nile virus secretion and morphology as a target for viral attenuation. J Virol 94(12):e00086–e00020. https://doi.org/10.1128/JVI.00086-20

8. Delbue S, Ferrante P, Mariotto S, Zanusso G et al (2014) Review of West Nile virus epidemiology in Italy and report of a case of West Nile virus encephalitis. J Neurovirol 20(5):437–441. https://doi.org/10.1007/s13365-014-0276-0

9. Setoh YX, Prow NA, Hobson-Peters J, Lobigs M et al (2012) Identification of residues in West Nile virus pre-membrane protein that influence viral particle secretion and virulence. J Gen Virol 93(Pt 9):1965–1975. https://doi.org/10.1099/vir.0.044453-0

10. Vidaña B, Busquets N, Napp S, Pérez-Ramírez E et al (2020) The role of birds of prey in West Nile virus epidemiology. Vaccines (Basel) 8(3):550. https://doi.org/10.3390/vaccines8030550

11. Barzon L, Pacenti M, Franchin E, Lavezzo E et al (2013) Whole genome sequencing and phylogenetic analysis of West Nile virus lineage 1 and lineage 2 from human cases of infection, Italy, August 2013. Euro Surveill 18(38):20591. https://doi.org/10.2807/1560-7917.es2013.18.38.20591. Erratum in: Euro Surveill 2013;18(40):pii/20597

12. Schvartz G, Farnoushi Y, Berkowitz A, Edery N et al (2020) Molecular characterization of the re-emerging West Nile virus in avian species and equids in Israel, 2018, and pathological description of the disease. Parasit Vectors 13(1):528. https://doi.org/10.1186/s13071-020-04399-2

Chapter 14

Mosquito Surveillance for West Nile Virus

Donald A. Yee, Ary Faraji, and Ilia Rochlin

Abstract

Identifying the mosquitoes responsible for transmitting human disease-causing pathogens is of critical importance for effective control of mosquito-borne outbreaks. West Nile virus is often transferred by adult female mosquitoes in the genus *Culex*, which deposit eggs in a variety of aquatic habitats throughout the world. Herein we describe several methodological approaches to monitor these species in nature, as well as offering details for data collection and analysis.

Key words *Aedes*, Mosquito adult, *Culex*, Modeling, Surveillance, Traps, Vector control

1 Introduction

West Nile virus (WNV) belongs to the Japanese encephalitis virus serogroup, is in the genus *Flavivirus* in the family *Flaviviridae*, and is wide-spread throughout North America and across the globe [1]. First described in 1937 from Uganda, it was detected in 1999 in New York and is now arguably the most widely studied mosquito-borne pathogen in the USA [2]. Infections of WNV in humans are often asymptomatic (80%) or produce mild, flu-like symptoms (~20%); however, in <1% of cases, infected individuals can experience severe neurological effects including memory loss, as well as death [1, 2]. Between 1999 and 2020, a total of 52,532 US WNV cases were reported, with 2456 deaths [3]. Worldwide cases are more difficult to estimate; however, the pathogen is ubiquitous and has caused outbreaks on most continents, including Africa, Asia, and Europe [1, 2].

The virus circulates among bird populations, especially among those in the Family Corvidae (although other species are affected); however, humans and other mammals are considered to be incidental or dead-end hosts, as viremia never reaches levels high enough to allow for transmission by mosquitoes [2]. Worldwide, the virus

Fengwei Bai (ed.), *West Nile Virus: Methods and Protocols*,
Methods in Molecular Biology, vol. 2585, https://doi.org/10.1007/978-1-0716-2760-0_14,

primary or enzootic vectors are all in the genus *Culex* (e.g., *Cx. modestus, Cx. pipiens, Cx. quinquefasciatus, Cx. tarsalis*) [4–7]. Several mosquito species, mostly from the genus *Aedes*, have also been implicated in potential transmission to humans and domestic animals as bridge vectors [2, 5]. Our current understanding of the vectors implicated in WNV transmission is likely incomplete, especially for those species involved in the epizootic transmission cycle. Surveillance of WNV vectors is thus critical to filling in knowledge gaps and preventing transmission to humans.

All stages of the mosquito life cycle, including eggs, larvae, and adults, are vital for our understanding of pathogen transmission [1]. Spatial and temporal distributions of vectors can lead to predictions about the underlying causes of outbreaks as well as to better inform control efforts. Surveillance includes several components, most commonly the collection and identification of adult and larval mosquitoes and testing of adult pools for the presence and frequency of pathogen infection [8]. The types of surveillance methods employed have the potential to affect the usefulness of the data, as there is no one "magic bullet" methodology that can answer every question related to vector-borne diseases. Thus, it is most useful to consider the research question or hypothesis and use the most relevant methodologies employed in a systematic way. It is also important to consider how such data can be shared broadly, so that both geographic and temporal trends can better inform limited vector control and suppression efforts. This chapter considers practical aspects of adult mosquito surveillance methodologies and ways in which data may be processed, stored, and shared to maximize our ability to control WNV transmission and safeguard public health.

2   Materials

2.1   CDC Light Trap (Figs. 1 and 2)

1. CDC light trap: purchased from a supplier or 3D printed (*see* **Note 1**).

2. Collection net or cup.

3. Six-volt rechargeable sealed lead-acid battery or D cell batteries (if a reliable AC source is not available for stationary traps).

4. Battery charger adapted for multiple batteries.

5. Source of carbon dioxide: dry ice in a 2- to 5-L dispenser/bag or bottled CO_2.

2.2   Gravid Water Infusion

1. Large container for making gravid water infused with organic material such as large trashcan (120 L) or more with tight-fitting lid.

Fig. 1 An example of a CDC light trap for collection of adult host-seeking female mosquitoes. Fully assembled unit in the field (left): (1) Trap body containing the fan. (2) Collection cup. (3) Stand. (4) Six-volt rechargeable lead acid battery. (5) Storage box with CO_2 bottle inside. *Insets*: (**a**) Up-close of the trap body. Wires supply power to the fan inside. Transparent tubing leads to the CO_2 bottle inside the storage box on the ground. (**b**) Trap body with the collection cup attached. Mosquitoes are drawn by the fan into the collection cup, which is taken off the unit while the fan is operating and the net is closed quickly to prevent escapees. (**c**) CO_2 bottle inside the storage box on the ground set up for the trap deployment. (**d**) Trap components that are transported to the trapping site stowed inside the storage box

2. Infusion material: hay, grass clippings, rabbit chow, fish emulsion, and manure.

3. Jerry cans (10 L) or plastic jugs for carrying the gravid water to the field sites.

2.3 Specimen Processing, Identification, and Storage

1. A cooler with ice or ice packs to store field-collected specimens.

2. Dissecting (stereo) microscope 10×–40× magnification.

3. Laboratory chill table.

4. Forceps: fine point, stainless steel, or epoxy.

5. Vials: Eppendorf or polypropylene snap cap rated for −80 °C.

6. Identification guide with appropriate morphological keys.

Fig. 2 Alternative trap designs. (**a**) CDC light trap with dry ice cooler with holes drilled around the bottom to supply CO_2 (1), trap body cover (2) to protect from rain, and a net instead of a collection cup for capturing mosquito specimens (3). A net may not be as convenient for transportation as a plastic collection cup. The trap is hung from a tree branch without the need for a stand. (**b**) CDC gravid trap, upright design. The net (1) is positioned on top of a tube containing the fan (2), which is hold in the upright position above the pan with gravid water (3). There are several disadvantages of this trap design. Mosquitoes and the trap battery are exposed to the elements (notice the cap covering the net). Furthermore, mosquitoes pass through the fan and into the collection net resulting in damaged specimens that are harder to identify and creating higher risk of contamination

7. Optional: equipment for molecular species identification by PCR.

8. An ultra-low temperature freezer (-80 °C) for specimen storage.

2.4 Data Processing and Analysis

1. Computer or tablet for record keeping and data analysis.

2. Printed spreadsheets or tablets for data recording.

3. Database: private or public such as VectorSurv.

4. Statistical/graphical software for data analysis: R, QGIS, and MS Excel.

3 Methods

3.1 Trap Site Selection

1. Traps should be placed in areas near potential mosquito larval habitat.

2. In rural areas, farms, especially those with waterfowl, can produce large number of *Culex* mosquitoes.

3. In urban areas, where vandalism is of concern, traps should be placed in secured locations, e.g., groundwater recharge basins, wastewater processing facilities, firehouses, or private homes.

4. Traps should be placed in a vegetated area preferably with tree or bush cover, in a shade and protected from strong winds.

5. If no suitable perch for CDC light trap is found, a simple 1.5-m pole (Fig. 1) can be used (e.g., Shepherds hook).

6. CDC light and gravid traps can be used in pairs with about 10-m distance between the two traps.

7. Established permanent trap sites are preferable for long-term data collection and surveillance programs.

8. Temporary sites are best used for pilot projects, to maximize mosquito collections over short time periods or to monitor new outbreaks.

3.2 CDC Light Trap (Fig. 1)

1. Ensure that the batteries are fully charged.

2. Ensure that you have sufficient amount of dry ice (*see* **Note 2**) or fully filled CO_2 bottles.

3. Use a trap cap or cover if rain is expected.

4. Assemble the trap at the field site.

5. Hang the trap from a tree branch or a pole at approximately 1.5 m above the ground.

6. Ensure that the fan is running and the trap light is on (*see* **Note 3**).

7. Deploy the trap in the late afternoon before dusk; collect in the morning after dawn (*see* **Note 4**).

3.3 CDC Gravid Trap (Fig. 3)

1. Ensure that the batteries are fully charged.

2. Fill the carrying jerry can with fresh gravid water (*see* **Note 5**).

3. The pan that supports the trap is filled with approximately 4 L of gravid water infusion.

4. Trap is placed at the field site before dusk and is removed the following morning.

5. Pan can be left in the field if covered with lid secured by a stone.

6. Old infusion can be used for no more than 7 consecutive days if no mosquito larvae are found inside the pan.

7. For routine surveillance, new infusion should be made and used weekly.

8. Box type where the mosquitoes don't go through the fan is preferable for obtaining undamaged specimens [9].

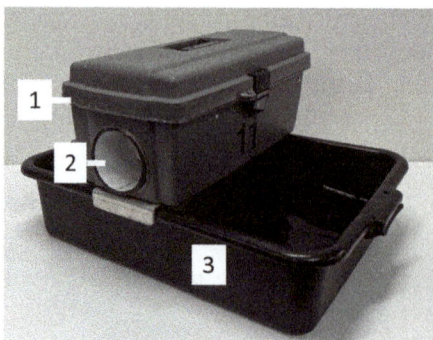

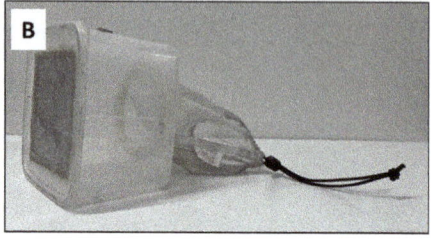

Fig. 3 A CDC gravid trap for collection of egg carrying females of *Cx. pipiens* complex mosquitoes. Fully assembled unit (left): (1) Box trap body containing the fan, the battery, and the collection container. (2) Fan opening. (3) Plastic pan for gravid water that attracts female mosquitoes. The pan is filled approximately halfway to leave some space for the intake opening (facing down) to aspirate mosquitoes that approach the pan to lay eggs. *Insets*: (**a**) Inside the box trap body (1) collection container with the opening facing down, (2) fan, and (3) six-volt rechargeable lead-acid battery. Note that mosquitoes are aspirated into the collection container without passing through the fan resulting in undamaged specimens for identification. (**b**) Collection container with the mesh sleeve that is closed while the fan is running to prevent escapees

3.4 Specimen Processing, Identification, and Storage

1. Remove collection cup or net from the trap (*see* **Note 6**).

2. Put collections cups (*see* **Note 7**) in the cooler. The cooler should have a layer of old newspapers over ice packs to absorb moisture. The remaining dry ice from the trap could also be transferred to the cooler and covered with newspaper.

3. Maintain cold chain from this point on including using air conditioning in the vehicle and inside laboratory facilities.

4. Disassemble and remove the traps as described in Subheadings 3.2 and 3.3.

5. Deliver the specimens to the laboratory facilities. To anesthetize mosquitoes put more dry ice in the cooler, or briefly expose the cups to cold temperatures in the freezer.

6. Use chilling table to sort and identify the specimens (*see* **Note 8**).

7. Put up to 50 mosquito females of the same species from the same trap in one tube (*see* **Note 9**). If more than 50 had been collected, make multiple pools using separate tubes.

8. Store the tubes with specimens at −80 °C.

9. Record all the data on a spreadsheet.

3.5 Data Processing and Analysis

1. Initial data recording can be done either on paper, on tablets, or using cell phones.

2. All data should be entered or ported to a database that links mosquito sample information (e.g., location, date, time, species collected, notes) with WNV data (e.g., pool number, test results).

3. Private databases can be designed and maintained using MS Access or similar software. However, statewide or regional data hubs such as VectorSurv are preferable for long-term data storage, data integration, and ease of access.

4. Data can be further processed, visualized, and analyzed using MS Excel add-ons for infection rate calculations [10] and freeware such as geographic information systems (GIS) software QGIS and statistical software R (*see* **Note 10**).

5. A large suite of basic and advanced modeling tools is available with statistical software R including basic statistical techniques, mixed effects models, Bayesian methods, spatial analysis, and machine learning.

4 Discussion

The main purpose of WNV mosquito surveillance is to monitor viral activity in vector species to determine human risk and to assist with vector control decision-making processes [11]. This mode of surveillance is typically carried out by governmental mosquito control or public health agencies. Furthermore, the same methodologies are utilized to obtain isolates of WNV for molecular or genetic characterization. Although WNV can be passed from the female mosquito to her offspring through transovarial or transgenerational transmission, this is a rare occurrence [12], and the virus is not typically expected to be isolated from larval *Culex* mosquitoes in the field. Male mosquitoes do not blood feed as adults, and therefore, for all practical considerations, WNV surveillance is focused on capturing adult female mosquitoes in sufficient numbers to enable virus detection, isolation, and quantification [11].

Collecting relatively high numbers of female mosquitoes is necessary because WNV usually has comparatively low prevalence in mosquito populations [13]. For this reason, WNV is commonly detected not from a single female mosquito specimen, but from 10 to 100 specimens combined or "pooled" together for processing and pathogen isolation. These mosquito pools are composed of female mosquitoes of the same species. Accordingly, processing of field collections necessitates separating female mosquitoes from males and other bycatch (nontarget insects) followed by morphological identification of females to a species level using published keys. In some cases, different species, particularly those from the genus *Culex*, are indistinguishable based on morphological characteristics [14]. In those instances, molecular identification of individual specimens or genetic material isolated from mosquito pools may be required [15]. Molecular methods may also supplement morphological identification if the specimens were damaged during the collection or transport process.

Two different trap types are used for routine collections of WNV mosquito vectors targeting different mosquito populations: the Centers for Disease Control and Prevention (CDC) light traps supplemented by carbon dioxide (CO_2) and CDC gravid traps. CDC light traps (Fig. 1) are used to sample host-seeking mosquito species. These traps collect a wide variety of mosquito species in diverse environments, providing data on species composition and multiple WNV vector species [16–18]. However, this trap type also has significant limitations. If located in close proximity to mosquito larval habitats, most specimens collected in CDC light traps may be unfed or nulliparous females that have yet to blood feed and acquire the virus. Some mosquito species, including important WNV vectors such as *Cx. pipiens*, can be under sampled by the CDC light traps, resulting in low numbers of collected specimens. Emitted light can also attract numerous bycatch insects that may damage target species and make sample processing more difficult and time-consuming. For this reason, many surveillance programs remove the light source from the trap and rely on carbon dioxide as the only attractant. Finally, CDC light traps are not suitable for sampling diurnal container-inhabiting *Aedes* species such as *Ae. albopictus*, *Ae. aegypti*, and *Ae. japonicus*. Other more specialized trap types, such as the BG Sentinel or Mosquito Magnet with different sets of lures, are deployed to capture these mosquito species [19].

Alternatively, CDC gravid traps (Fig. 3) are a specialized collection device targeting gravid females of the *Cx. pipiens* complex, which are prevalent in peridomestic habitats [16–18, 20]. Gravid females are those that have obtained a blood meal and are carrying eggs ready for oviposition. Thus, CDC gravid traps have a much higher likelihood of capturing mosquitoes that have been infected with WNV and hence a higher propensity to detect or isolate the virus. However, due to the high selectivity of CDC gravid traps for *Cx. pipiens* complex mosquitoes, important mosquito vector species such as *Cx. tarsalis* may not be collected in sufficient numbers [13]. Therefore, CDC gravid traps are frequently deployed alongside CDC light traps in many jurisdictions. Gravid traps also require an oviposition attractant, specifically water infused with organic material such as hay, rabbit chow, or fish emulsion that mimic stagnant water present in larval habitats preferred by *Cx. pipiens* complex mosquitoes [21]. Various materials used for infusion may exhibit different attractiveness for local *Cx. pipiens* complex species [18], and gravid (also known as stagnant or "stink") water can be cumbersome to maintain and distribute. Care should be taken not to leave the trap container with gravid water exposed because, if left unattended, these can become significant mosquito production sources themselves. The attractiveness of the CDC gravid traps to mosquitoes also varies based on what competing natural larval habitats are found in the immediate vicinity of the trap. Regardless of the type of trap utilized for WNV mosquito surveillance, attention must be given to specific materials and methods within each jurisdiction, which may vary based on geography, species composition, logistics, and even economics.

We note that in remote sampling areas, researchers may not have access to bottled CO_2 or electricity to recharge batteries for traps. In such cases, small solar panels that can charge a battery during the day so that the trap may be operated during the nocturnal or night-time hours would be beneficial (Fig. 1). In addition, when CO_2 is not available, artificial CO_2 generators can be used from fermentation of sugar by yeast in a closed environment (e.g., Biogents BG-CO_2 Generator, Germany). Although neither of these approaches may be ideal, they may offer the only opportunity for collections in otherwise remote and inaccessible locations.

5   Notes

1. List of suppliers (USA): John W. Hock Co (https://www.johnwhock.com/), Clarke Mosquito Control (https://www.clarke.com/field-and-surveillance-tools). Contact one of the authors (AF) for 3D printer instructions.

2. Solid blocks of dry ice used in ice cream establishments are preferable to pellets. Dry ice blocks can be stored in −80 °C refrigerators for several days. Each approximately 30-kg block is split into 1–2 kg cubes using mallet and chisel if a soft bag is used. Plastic dispenser requires smaller pieces of the same total weight. Personal protection equipment should be utilized to protect the face, eye, and skin against dry ice shards during cutting process. Bags with dry ice are transported in a cooler.

3. If the trap is equipped with a photoswitch to be operated from dawn to dusk, cover the photoswitch with the palm of your hand to check that the trap is operational. Trap bulb can be removed to reduce non-mosquito bycatch.

4. Target mosquitoes are most active at dawn and dusk, so be certain the trap is deployed during those times of the day. When traps are used with optional photoswitch and a Gate system available from some suppliers, the CDC light trap can be run 3–4 days on one battery requiring only one visit per day to collect the sample and to replenish CO_2. This system can also eliminate late evening and early morning work.

5. The hay infusion is made by adding 0.5 kg of hay to approximately 100 L of tap water, and allowing the infusion to incubate for 5 days. The length of the incubation period is temperature dependent and may be shorter at higher temperatures or longer at lower temperatures. Use leak-proof jerry cans to carry gravid water: it has a very disagreeable odor if spilled inside the vehicle.

6. Be certain to close the net's opening quickly while taking it off the trap—mosquitoes can fly out and escape.

7. Plastic cups are preferable to nets because the mosquitoes inside are not easily damaged. Cups can be stacked up in a cooler. Nets have to be hung from a string (can be positioned between handles over the rear seat of the vehicle) using clothespins.

8. After the initial knockdown, it is advisable to transfer the mosquito specimens from each trap to a pre-labeled petri dish held together by a rubber band. If chilling table is not available, the lab should be kept cool and mosquitoes processed in small batches.

9. For species identification, regional or local keys are easier and simpler to use. For molecular species identification, a leg from each individual is removed and stored in a separate tube for DNA isolation.

10. GIS software QGIS (https://www.qgis.org), R software for statistical computing (https://www.r-project.org/). Software to calculate infection rates (MIR/MLE) is available from CDC (https://www.cdc.gov/westnile/resourcepages/mosqSurvSoft.html). Estimates of the infection rates are usually presented as the number of infected mosquitoes per 1000 tested. The minimum infection rate (MIR) is calculated: ([number of positive pools / total specimens tested] × 1000). The MIR uses the assumption that a positive pool contains only one infected mosquito and may be invalid when infection rates are high. MLE estimates do not require the assumption used in the MIR calculation. This program also includes calculation of confidence intervals which reflect, in part, the sample sizes used in the calculations. The confidence intervals (or any other uncertainty measure) are essential for interpreting the precision of the IR estimate.

11. We note that in remote sampling areas, researchers may not have access to bottled CO_2 or electricity to recharge batteries for traps. In such cases, small solar panels that can charge a battery during the day so that the trap may be operated during the nocturnal or night-time hours would be beneficial (see, e.g., Auto-Sensing 6- and 12-Volt 30-Wt Solar-Powered Lead-

Acid Battery Charger https://www.johnwhock.com/products/batteries-chargers/). In addition, when CO_2 is not available, artificial CO_2 generators can be used from fermentation of sugar by yeast in a closed environment (e.g., Biogents BG-CO_2 Generator, Germany, https://eu.biogents.com/bg-co2-generator/). Although neither of these approaches may be ideal, they may offer the only opportunity for collections in otherwise remote and inaccessible locations.

Acknowledgments

We thank Fengwei Bai for inviting us to participate in this publication.

References

1. Clements AN (2011) The biology of mosquitoes, volume 3: transmission of viruses and interactions with bacteria. CABI Publishing, New York, NY, USA

2. Kramer LD, Styer LM, Ebel GD (2008) A global perspective on the epidemiology of West Nile virus. Annu Rev Entomol 53:61–81. https://doi.org/10.1146/annurev.ento.53.103106.093258

3. CDC (2021) Final cumulative maps and data – West Nile Virus. https://www.cdc.gov/westnile/statsmaps/cumMapsData.html. Accessed 4 Feb 2022

4. Fonseca DM, Keyghobadi N, Malcolm CA et al (2004) Emerging vectors in the *Culex pipiens* Complex. Science 303:1535–1538. https://doi.org/10.1126/science.1094247

5. Rochlin I, Faraji A, Healy K et al (2019) West Nile virus mosquito vectors in North America. J Med Entomol 56:1475–1490. https://doi.org/10.1093/jme/tjz146

6. Farajollahi A, Fonseca DM, Kramer LD et al (2011) "Bird biting" mosquitoes and human disease: a review of the role of *Culex pipiens* complex mosquitoes in epidemiology. Infect Genet Evol 11:1577–1585. https://doi.org/10.1016/j.meegid.2011.08.013

7. Hubalek Z (2000) European experience with the West Nile virus ecology and epidemiology: could it be relevant for the New World? Viral Immunol 13:415–426

8. Gu W, Unnasch TR, Katholi CR et al (2008) Fundamental issues in mosquito surveillance for arboviral transmission. Trans R Soc Trop Med Hyg 102:817–822. https://doi.org/10.1016/j.trstmh.2008.03.019

9. Kesavaraju B, Kiyoguchi D, Dickson S (2011) Efficacy of gravid traps in trapping *Culex pipiens*. J Am Mosq Control Assoc 2011(10/25):320–322

10. (2021) Mosquito Surveillance Software | West Nile Virus | CDC. https://www.cdc.gov/westnile/resourcepages/mosqSurvSoft.html. Accessed 4 Feb 2022

11. CDC (2013) West Nile virus in the United States: guidelines for surveillance, prevention, and control

12. Anderson JF, Main AJ (2006) Importance of vertical and horizontal transmission of West Nile virus by *Culex pipiens* in the northeastern United States. J Infect Dis 194:1577–1579

13. Dunphy BM, Kovach KB, Gehrke EJ et al (2019) Long-term surveillance defines spatial and temporal patterns implicating *Culex tarsalis* as the primary vector of West Nile virus. Sci Rep 9:6637. https://doi.org/10.1038/s41598-019-43246-y

14. Harrington LC, Poulson RL (2008) Considerations for accurate identification of adult *Culex restuans* (Diptera: Culicidae) in field studies. J Med Entomol 45:1–8

15. Rochlin I, Santoriello MP, Mayer RT et al (2007) Improved high-throughput method for molecular identification of Culex mosquitoes. J Am Mosq Control Assoc 23:488–491

16. Reisen WK, Boyce K, Cummings RC et al (1999) Comparative effectiveness of three adult mosquito sampling methods in habitats

representative of four different biomes of California. J Am Mosq Control Assoc 15:24–31

17. McCardle PW, Webb RE, Norden BB et al (2004) Evaluation of five trapping systems for the surveillance of gravid mosquitoes in Prince Georges county, Maryland. J Am Mosq Control Assoc 20:254–260

18. Burkett DA, Kelly R, Porter CH et al (2004) Commercial mosquito trap and gravid trap oviposition media evaluation, Atlanta, Georgia. J Am Mosq Control Assoc 20:233–238

19. Rochlin I, Kawalkowski M, Ninivaggi DV (2016) Comparison of mosquito magnet and biogents sentinel traps for operational surveillance of container-inhabiting Aedes (Diptera: Culicidae) species. J Med Entomol 53:454–459. https://doi.org/10.1093/jme/tjv171

20. Reiter P, Jakob WL, Francy DB et al (1986) Evaluation of the CDC gravid trap for the surveillance of St. Louis encephalitis vectors in Memphis, Tennessee. J Am Mosq Control Assoc 2:209–211

21. McNamara TD, Healy K (2021) A comparison of Hay and fish emulsion-infused water as oviposition attractants for the CDC gravid trap. J Med Entomol. https://doi.org/10.1093/jme/tjab203

Chapter 15

Molecular Surveillance of West Nile Virus in Mosquitoes and Sentinel Chickens

Steven T. Peper

Abstract

Arboviral surveillance is a critical step in any effective mosquito control program. Surveillance aids in the early detection of pathogen transmission as well as establishes a baseline of transmission activity. Two of the most practical forms of arboviral surveillance is through the use of testing mosquito pools for the presence of pathogen and screening sentinel chickens for pathogen exposure. This chapter describes the process for each of these methods for West Nile virus.

Key words ELISA, Molecular techniques, Mosquitoes, PCR, Sentinel chickens, West Nile virus

1 Introduction

Arboviruses such as chikungunya virus (CHIKV), dengue virus (DENV), St. Louis encephalitis virus (SLEV), West Nile virus (WNV), yellow fever virus (YFV), Zika virus (ZIKV), and others pose a continued threat to public health on a global scale [1]. The management of mosquito populations is one of the main line of defenses against arboviruses and can be enhanced through surveillance efforts [1]. As such, arboviral surveillance is a critical function of any effective mosquito control program, whether this surveillance is done through in-house capabilities or outsourced through external laboratories. Commonly used surveillance technique include monitoring disease cases in humans and wild animals, sentinel animal surveillance, and mosquito surveillance [1]. A comprehensive surveillance strategy is critical for the early detection of arboviral transmission but also helps set a baseline for arboviral activity in a community [2]. Surveillance also helps document the distribution of arboviral activity which aids in the determination of risk for human and animal populations [3].

Fengwei Bai (ed.), *West Nile Virus: Methods and Protocols*,
Methods in Molecular Biology, vol. 2585, https://doi.org/10.1007/978-1-0716-2760-0_15,

Animal surveillance has its merits as it can often serve as an early warning for human disease cases [4]. As an example, reports of dead birds played a critical role in the identification of the WNV outbreak in New York in 1999 [5]. Unfortunately, as a passive form of surveillance, human and animal surveillance has its limitations [1] and, therefore, this chapter will focus on the molecular surveillance through mosquito pool and sentinel chicken testing.

2 Materials

2.1 Arboviral Surveillance in Mosquito Pools

1. Mosquito traps.
2. 1 × chloride-Tris-EDTA (STE) buffer: 10 Mm of Tris/Tris–HCI, 1 mM of EDTA, 100 mM of NaCl.
3. Grinding balls, stainless steel, 5/32″ (*see* **Note 1**).
4. 2-mL Microcentrifuge tubes.
5. Bead Mill Homogenator.
6. Microcentrifuge (6000–20,000 × g).
7. 100% Ethanol.
8. QIAamp Viral RNA Mini Kit (*see* **Note 2**).
   - (a) Buffer AVE.
   - (b) Carrier RNA (*see* **Note 3**).
   - (c) Buffer AVL, viral lysis buffer.
   - (d) Buffer AW1, wash buffer 1 concentrate (*see* **Note 4**).
   - (e) Buffer AW2, wash buffer 2 concentrate (*see* **Note 4**).
   - (f) QIAamp mini spin columns.
   - (g) Wash tubes.
9. TaqPath 1-Step Multiplex Master Mix (Applied Biosystems TaqPath 1-Step Multiplex Master Mix, Cat #: A28526).
10. Primers and probes (*see* **Note 5**).
    - (a) West Nile virus [6].
      - (i) Forward: TCA GCG ATC TCT CCA CCA AAG.
      - (ii) Reverse: GGG TCA GCA CGT TTG TCA TTG.
      - (iii) Probe: TGC CCG ACC ATG GGA GAA GCT C.
11. Nuclease-free water.
12. Positive control (*see* **Note 6**).
13. PCR Plate.
14. PCR film cover.
15. Vortex.
16. PCR plate centrifuge.
17. qRT-PCR machine (*see* **Note 7**).

2.2 Detection of West Nile Virus Exposure Using Sentinel Chickens

1. Sentinel chicken flocks (*see* **Note 8**).

2. 23- or 25-Gauge needle and a 3-cc syringe.

3. 3-mL Serum separator tubes.

4. Centrifuge to fit 3-mL serum separator tubes.

5. Deionized (DI) or reverse osmosis (RO) water.

6. ID Screen® West Nile Competition Multi-species ELISA Kit (*see* **Note 9**).

    (a) Pre-coated ELISA plate.

    (b) Dilution buffer.

    (c) Positive control.

    (d) Negative control.

    (e) Wash solution 20× (*see* **Note 10**).

    (f) Concentrated conjugate (*see* **Note 11**).

    (g) Substrate solution.

    (h) Stop solution.

7. Microplate washer (*see* **Note 12**).

8. Microplate OD reader.

3 Methods

3.1 Arboviral Surveillance in Mosquito Pools Using qRT-PCR Tri-Plex Assay

Female mosquitoes can be collected via a variety of trap types (i.e., gravid traps, CDC mini light traps, EVS CO_2-baited traps, etc. [7]) for qRT-PCR analysis. Once collected, mosquitoes should be euthanized by placing in a freezer (≤ -20 °C) for ≥ 1 h. Once euthanized, mosquitoes should be kept cold during the identification and pooling process as to preserve any potential viral RNA. After identification, mosquitoes should be pooled (≤ 50 mosquitoes per pool) by species and location and placed in 2-mL microcentrifuge tubes.

3.1.1 Extraction of RNA from Mosquito Pools Using QIAamp Viral RNA Mini Kit

1. After collection and pooling of mosquito pools, add 250 μL of 1× STE buffer into all tubes containing <20 mosquitoes, and add 500 μL of 1× STE buffer into all tubes containing ≥20 mosquitoes.

2. Add one stainless steel grinding ball to each tube and close the cap.

3. Using the bead mill homogenator, grind all mosquitoes for 30 s or until emulsified.

4. Centrifuge all emulsified mosquito tubes at 20,000 × g for 10 min.

5. Add 5.6 μL of carrier RNA and 560 μL of buffer AVL into an empty 2-mL microcentrifuge tube (*see* **Note 13**).

6. Pipette 140 µL of the supernatant from centrifuged mosquito tube into the tube containing carrier RNA and AVL.

7. Pulse vortex tubes for ≥15 s, and incubate at room temperature (15–25 °C) for 10 min.

8. After incubation, add 560 µL of 100% ethanol to the sample.

9. Pulse vortex for ≥15 s, and then briefly centrifuge to remove drops from the inside of the lid (*see* **Note 14**).

10. Carefully apply 630 µL of the solution from step 8 (supernatant/buffer AVL/carrier RNA/100% ethanol) to the QIAamp Mini Spin Column (in a wash tube (WT)).

11. Close the cap and centrifuge at approximately 6000 × g for 2 min.

12. Place the QIAamp Mini Spin Column into a clean WT, and discard the wash tube containing the filtrate.

13. Carefully open the QIAamp Mini Spin Column and repeat **steps 10–12**.

14. Carefully open the QIAamp Mini Spin Column, and add 500 µL of diluted buffer AW1 (*see* **Note 4**).

15. Close the cap and centrifuge at approximately 6000 × g for 2 min.

16. Place the QIAamp Mini Spin Column into a clean WT, and discard the wash tube containing the filtrate.

17. Carefully open the QIAamp Mini Spin Column and add 500 µL of diluted buffer AW2 (*see* **Note 4**).

18. Close the cap and centrifuge at 20,000 × g for 4 min.

19. Place the QIAamp Mini Spin Column into a clean WT, and discard the wash tube containing the filtrate.

20. Centrifuge again at 20,000 × g for 1 min.

21. Place the QIAamp Mini Spin Column into a clean WT, and discard the WT containing the filtrate (*see* **Note 15**).

22. Carefully open the QIAamp Mini Spin Column and add 60 µL of buffer AVE directly onto the filter (*see* **Note 16**).

23. Close the cap and incubate at room temperature for 1 min.

24. Centrifuge at 20,000 × g for 2 min.

25. Transfer the eluate from the WT into a clean 2-mL microcentrifuge tube.

26. Label tube with "RNA Isolate" and any specific identification number.

27. Store RNA isolate at −20 °C until ready for analysis (stable for up to a year). Long-term storage should be at −80 °C.

Table 1
Matrix to generate a master mix for the qRT-PCR assay

Master mix	Volume	$n + 3$	Total
RNase-free H_2O	8.75		0
TaqPath master mix	5		0
WN F. primer 10 μM	0.5		0
WN R. primer 10 μM	0.5		0
WN probe 10 μM	0.25		0
RNA	5	–	–
Total volume	20	–	0

3.1.2 qRT-PCR Tri-Plex Assay

1. Prepare a PCR master mix (Table 1) by combining nuclease-free water, TaqPath 1-Step Multiplex Master Mix, primers, and probe into a single microcentrifuge tube (*see* **Note 17**).

2. "n + 3" from Table 1 (*see* **Note 18**) represents the number of samples plus the positive control, negative control, and one extra for pipette error (*see* **Note 19**).

3. Prepare a plate map for your 96-well PCR plate (Table 2).

4. Pipette 15 μL of the prepared master mix from step 1 into each well of the PCR plate that will have samples or controls (*see* **Note 20**).

5. Pipette 5 μL of each sample (RNA isolate from Subheading 3.1.1) and control into their corresponding wells.

6. Seal the plate with lids or clear adhesive cover, vortex the plate, and centrifuge to bring the contents to the bottom of the wells.

7. Put the plate in the qRT-PCR machine with the following cycle conditions:

   (a) Lid: 105 °C.

   (b) Step 1. 25 °C for 2 min.

   (c) Step 2. 53 °C for 10 min.

   (d) Step 3. 95 °C for 2 min.

   (e) Step 4. 95 °C for 15 s.

   (f) Step 5. 60 °C for 1 min.

   (g) Plate read.

   (h) Go to step 4 (e) and repeat 39 times.

8. At the completion of the PCR run, analyze results by setting the specific critical threshold (CT) cutoff values for each primer/probe sets.

Table 2
Suggested layout for the qRT-PCR plate

	1	2	3	4	5	6	7	8	9	10	11	12
A	Sample 1	Sample 2	Sample 3	Sample 4	Sample 5	Sample 6	Sample 7	Sample 8	Sample 9	Sample 10	Sample 11	Sample 12
B	Etc.											
C												
D												
E												
F												
G												
H	WNV Pos											NTC

9. Any CT value ≥ 38 should be considered negative.

10. Any CT value <38 should be considered positive (*see* **Note 21**).

3.2 Detection of West Nile Virus Exposure Using Sentinel Chickens

3.2.1 Collection of Blood/ Serum from Sentinel Chickens

1. Generally, sentinel chicken flocks include six individual female birds per flock (*see* **Note 22**).

2. Prior to establishing sentinel chicken flocks in the field, each chicken should be bled via the brachial veins in the wing using 23- or 25-gauge needles with 3-cc syringes [8, 9] and tested to establish a WNV antibody-negative baseline.

3. Once a week after chickens are placed in the field, they should be bled for testing.

4. Transfer blood from the syringe into the serum separator tubes (SST) (*see* **Note 23**).

5. After 30–60 min to allow for coagulation, SST should be centrifuged at $\geq 10,000 \times$ g for 10–20 min.

6. Serum can then be aliquoted from the SST, and stored or used directly for analysis (*see* **Note 24**).

3.2.2 Innovative Diagnostics ELISA Kit

1. Bring all reagents to room temperature prior to running ELISA and reagent preparation.

2. Prepare a 1× wash solution by diluting the kit-provided 20× wash concentration using distilled or deionized water (*see* **Note 25**).

3. Prepare a 1× conjugate by diluting the kit provided 10× concentrated conjugate using the kit provided dilution buffer (*see* **Note 26**).

4. Using a 96-well microplate (not provided by the kit), add 100 µL of kit-provided positive control into wells A1 and B1 (*see* **Note 27**).

5. Add 100 µL of kit-provided negative control into wells C1 and D1.

6. Add 100 µL of individual sentinel chicken samples into the remaining wells (one well per sample).

7. Add 100 µL of kit-provided dilution buffer into all control and sample wells.

8. Using a multichannel pipette, transfer 100 µL of diluted controls/samples into the corresponding kit-provided pre-coated microplate wells.

9. Incubate pre-coated plate for 90 min ($\pm$ 6 min) at 21 °C ($\pm$ 5 °C).

10. After incubation, aspirate contents of wells, and wash each well three times with approximately 300 µL of pre-prepared 1× wash solution (*see* **Note 28**).

11. Using a multichannel pipette, add 100 μL of the pre-prepared 1× conjugate to each well.

12. Incubate for 30 min (±3 min) at 21 °C (±5 °C).

13. After incubation, aspirate contents of wells, and wash each well three times with approximately 300 μL of pre-prepared 1× wash solution.

14. Using a multichannel pipette, add 100 μL of the kit-provided substrate solution to each well.

15. Incubate for 15 min (± 2 min) at 21 °C (± 5 °C) in the dark (*see* **Note 29**).

16. Add 100 μL of kit-provided stop solution to each well to stop reaction (*see* **Note 30**).

17. Read the plate OD at 450 nm (*see* **Note 31**).

18. ELISA validation:

    (a) The test is valid if the mean value of the negative control OD is greater than 0.700.

    (b) The test is valid if the mean value of the positive control OD is less than 30% of the mean negative control OD ($OD_{PC} / OD_{NC} < 0.30$).

19. Interpretation of ELISA results:

    (a) Calculate the S/N percentage (S/N%) for each sample: $S/N\% = (O.D._{Sample} / O.D._{NC}) * 100$.

    (b) S/N% $\leq$ 40% are considered positive.

    (c) S/N% > 50% are considered negative.

    (d) S/N% > 40% and $\leq$50% are considered doubtful.

4   Notes

1. Different-sized and quantity of grinding balls may be used based on preference.

2. Several commercial extraction methods/kits are available. Anastasia Mosquito Control District (AMCD) in St. Augustine, Florida, uses the QIAamp Viral RNA Mini Kit (QIAGEN, Cat #: 52906) to extract RNA from mosquito pools and serum samples which can be done with an automated extraction system, such as the QIAcube (QIAGEN) or by hand.

3. The lyophilized carrier RNA needs to be reconstituted at first use. Add 1 μL of AVE for every μg of lyophilized carrier RNA. Store at −20 °C.

4. Buffers AW1 and AW2 need to be diluted before first use. Add 100% ethanol to each bottle (specific volume is listed on specific bottles) for a final working concentration.

5. Several commercial companies are available to order primer/probes. AMCD orders from Integrated DNA Technologies, Inc. (IDT DNA). The WNV probe is designed with a FAM modification on the 5' end and BHQ-1 on the 3' end. You can also use this primer/probe with EEEV and SLEV primers/probes to form a triplex assay is desired.

   (a) Eastern equine encephalitis virus [10].

      (i) Forward: ACA CCG CAC CCT GAT TTT ACA.

      (ii) Reverse: CTT CCA AGT GAC CTG GTC GTC.

      (iii) Probe: TGC ACC CGG ACC ATC GGA CCT.

   (b) St. Louis encephalitis virus [11].

      (i) Forward: CTG GCT GTC GGA GGG ATT CT.

      (ii) Reverse: TAG GTC AAT TGC ACA TCC CG.

      (iii) Probe: TCT GGC GAC CAG CGT GCA AGC CG.

   The EEEV probe is designed with a HEX modification on the 5' end and BHQ-1 on the 3' end. The SLEV probe is designed with a CY5 modification on the 5' end and BHQ-2 on the 3' end.

6. West Nile virus (West Nile RT-PCR, "West Nile RNA lysate"), Eastern equine encephalitis virus (eastern equine encephalitis RT-PCR, "SINV/EEEV FTA Card"), and St. Louis encephalitis virus (St. Louis RT-PCR, "St. Louis encephalitis FTA Card") positive controls were provided by the Centers for Disease Control and Prevention (CDC), National Center for Emerging and Zoonotic Infectious Diseases (NCEZID), and Division of Vector-Borne Diseases (DVBD) in Fort Collins, Colorado. An account must be requested and created for ordering purposes through the CDC DVBD, Arboviral Diseases Branch, Arbovirus Reference Collection Reagent Ordering system (https://wwwn.cdc.gov/ReagentOrder/Default.aspx).

7. Several real-time PCR machines are available on the market. Anastasia Mosquito Control District uses the CFX96 optical reaction module with the C1000 touch thermal cycler (Bio-Rad Laboratories, Inc.).

8. The state of Florida recommends the use of Leghorn, Barred Rock, Rhode Island, or Minorcan chickens for sentinel use. Only female birds are recommended as the risk of cocks crowing may annoy residents in the area. Chickens should be 10–12 weeks of age prior to placement in the field [9].

9. Innovative Diagnostics offers a commercial competitive ELISA kit (ID Screen® West Nile Competition Multi-species) for the serological detection of West Nile virus exposure in horse and avian species as well as other species. The kit is offered as a

single plate (Cat #: WNC-1P) or a two-plate bundle (Cat #: WNC-2P). This kit is designed to detect antibodies against WNV envelope protein (pr-E). Innovative Diagnostics recognizes that the monoclonal antibody that is used in this kit is known to cross-react with other Japanese encephalitis viruses as well as the tick-borne encephalitis virus. Innovative Diagnostics is based in France and requires an import permit (United States Veterinary Biological Product Permit Transit Shipment Only) from the United States Department of Agriculture, Animal and Plant Health Inspection Service, Veterinary Services, Center for Veterinary Biologics (APHIS Form 2005). A Toxic Substance Control Act (TSCA) Certification for FedEx Express shipments must also be completed and filed with FedEx.

10. Wash solution should be diluted with DI or RO water.

11. Concentrated conjugate should be diluted with kit-provided dilution buffer prior to use.

12. DI or RO water should be used for the rinsing of the microplate washer. DI water is preferable as it would reduce the residue buildup on the equipment. The plate washer should be rinsed before, in-between, and after each use.

13. This reagent can be prepared in a larger quantity by multiplying each reagent by the number of samples to be tested (i.e., for ten samples mix 5600 μL of buffer AVL and 56 μL RNA carrier).

14. Do not centrifuge for too long as precipitated RNA may pellet at the bottom of the tube and not adequately be collected in future steps.

15. Original protocol calls for putting the spin column directly into a 2-mL microcentrifuge tube prior to centrifugation. Caps of these microcentrifuge tubes will often break off during centrifugation. Using an extra WT in this step and transferring to the 2-mL microcentrifuge tube after centrifugation solves this problem.

16. Elution with 60 μL of buffer AVE will elute at least 90% of the viral RNA from the spin column. Performing a double elution (two elutions with 40 μL of buffer AVE each) will increase yield up to 10%.

17. If used as a triplex assay, adjust the master mix formula as found in Table 3.

18. If used as a triplex assay, then note the "n + 5" from the Table in **Note 17**. "n + 5" now represents the number of samples plus the three positive controls, one negative control, and one extra for pipette error.

19. For a small number of samples (i.e., 1–24), one extra for pipetting error is often used. However, for larger number of samples, adding 10% of the total number of samples/controls

Table 3
Matrix to generate a master mix for a triplex qRT-PCR assay

Master mix	Volume	n + 5	Total
RNase-free H_2O	6.25		0
TaqPath master mix	5		0
WN F. primer 10 µM	0.5		0
WN R. primer 10 µM	0.5		0
EEE F. primer 10 µM	0.5		0
EEE R. primer 10 µM	0.5		0
SLE F. primer 10 µM	0.5		0
SLE R. primer 10 µM	0.5		0
WN probe 10 µM	0.25		0
EEE probe 10 µM	0.25		0
SLE probe 10 µM	0.25		0
RNA	5	–	–
Total volume	20	–	0

is often used for calculating master mix preparation to account for pipetting error. For example, instead of the "n + 3" master mix calculation, the new equation would be "(n + 2) * 1.1." As this new method of accounting for pipette error will increase the waste of reagents, if you are confident with the calibration and consistency of your pipettes, just adding 1–2 extra for pipetting error is often acceptable for number of samples over 24.

20. PCR plate should be kept cool during this step and until placement into the PCR machine by using a PCR plate cooler. Many options are available from a variety of companies. Anastasia Mosquito Control District uses one from Eppendorf (PCR-Cooler, Cat #: 022510525).

21. With high CT values (32–38), it is often advised to run those samples again using a secondary primer set that targets another region of the desired target pathogen for confirmation. Anastasia Mosquito Control Association uses the following primer set for the confirmation of WNV:

   (a) West Nile virus [6]:

   (i) Forward: CAG ACC ACG CTA CGG CG.

   (ii) Reverse: CTA GGG CCG CGT GGG.

   (iii) Probe: TCT GCG GAG AGT GCA GTC TGC GAT.

(b) The WNV probe for this primer/probe set is designed with a ROX modification on the 5′ end and BHQ-2 on the 3′ end.

22. The number of flocks utilized is dependent upon the size of your surveillance area. As a reference, the Anastasia Mosquito Control District of St. Johns County, Florida, has ten sentinel chicken flocks throughout the county which is 821 square miles in size. Man power and resource limitations also often determine the number of flocks used.

23. Anastasia Mosquito Control District uses SSTs from BD (Becton, Dickinson and Company, Cat #: 367981). After blood has been transferred to the SST, shake each tube several times to help activate the coagulation process. For a more in-depth process of how to bleed sentinel chickens, please refer to the Mosquito-Borne Disease Guidebook [9].

24. If after the initial centrifugation there isn't great serum separation, shake the tube and re-centrifuge. Typically, 2 mL of blood will produce about 1 mL of serum. However, more or less serum could be separated depending on the health of the individual bird.

25. 1× wash solution can be made in bulk (i.e., mixing 60 mL of 20× concentrate that is provided with the kit with 1140 mL of water) and stored in the fridge in-between uses.

26. To account for pipetting error, prepare extra 1× conjugate with each run (i.e., mix 100 μL of 10× concentrated conjugate with 900 μL of dilution buffer for each strip of eight wells being used).

27. 50 μL of each for **steps 1–4** can be used if there is not enough serum from your samples. 100 μL is used to account for any pipetting error.

28. Anastasia Mosquito Control District uses the Model 1575 Immunowash Microplate Washer (Bio-Rad) for plate washing.

29. Covering the plate with a piece of aluminum foil works well to block the light from the reagents.

30. Gently tap the sides of the plate if needed to fully mix the two reagents. Any blue coloration should turn yellow after addition of the stop solution.

31. Anastasia Mosquito Control District uses the iMark Microplate Absorbance Reader (Bio-Rad) for OD determination. Read the plate within 15 min of adding the stop solution for best results.

References

1. Ramirez AL, van den Hurk AF, Meyer DB, Ritchie SA (2018) Searching for the proverbial needle in a haystack: advances in mosquito-borne arbovirus surveillance. Parasit Vectors 11:320

2. Van den Hurk AF, Hall-Mendelin S, Johansen CA, Warrilow D, Ritchie SA (2012) Evolution of mosquito-based arbovirus surveillance systems in Australia. J Biomed Biotechnol 2012: Article ID:8325659

3. Langevin SA, Bunning M, Davis B, Komar N (2001) Experimental infection of chickens as candidate sentinels for West Nile virus. Emerg Infect Dis 7:726–729

4. Eidson M, Kramer L, Stone W, Hagiwara Y, Schmit K, New York State West Nile virus Avian Surveillance Team (2001) Dead bird surveillance as an early warning system for West Nile virus. Emerg Infect Dis 7:631–635

5. Lanciotti RS, Roehrig JT, Deubel V, Smith J, Parker M, Steele K, Crise B, Volpe KE, Crabtree MB, Scherret JH, Hall RA, Mackenzie JS, Cropp CB, Panigraphy B, Ostlund E, Malkinson M, Banet C, Weissman J, Komar N, Savage HM, Stone W, Mcnamara T, Gubler DJ (1999) Origin of the West Nile virus responsible for an outbreak of encephalitis in the northeastern United States. Science 286:2333–2337

6. Lambert AJ, Martin DA, Lanciotti RS (2003) Detection of North American eastern and western equine encephalitis viruses by nucleic acid amplification assays. J Clin Microbiol 41: 379–385

7. Lanciotti RS, Kerst AJ, Nasci RS, Godsey MS, Mitchell CJ, Savage HM, Komar N, Panella NA, Allen BC, Volpe KE, Davis BS, Roehrig JT (2000) Rapid detection of West Nile virus from human clinical specimens, field-collected mosquitoes, and avian samples by a TaqMan reverse transcriptase-PCR assay. J Clin Microbiol 38:4066–4071

8. Johnson AJ, Langevin S, Wolff KL, Komar N (2003) Detection of anti-West Nile virus immunoglobulin M in chicken serum by enzyme-linked immunosorbent assay. J Clin Microbiol 41:2002–2007

9. Florida Department of Health (2021) Mosquito-Borne Disease Guidebook. http://www.floridahealth.gov/diseases-and-conditions/mosquito-borne-diseases/guidebook.html?utm_source=frywj_floridahealth.gov&utm_medium=referral&utm_campaign=fdoh_az_index. Accessed 1 Dec 2021

10. Lanciotti RS, Kerst AJ (2001) Nucleic acid sequence-based amplification assays for rapid detection of West Nile and St. Louis encephalitis viruses. J Clin Microbiol 39:4506–4513

11. Kading RC, Cohnstaedt LW, Fall K, Hamer GL (2020) Emergence of arboviruses in the United States: the boom and bust of funding, innovation, and capacity. Trop Med Infect Dis 5:96

Chapter 16

Statistical Tools for West Nile Virus Disease Analysis

Matthew J. Ward, Meytar Sorek-Hamer, Krishna Karthik Vemuri, and Nicholas B. DeFelice

Abstract

West Nile virus (WNV) is the most widespread arbovirus in the world and endemic to much of the United States. Its range continues to expand as land use patterns change, creating more habitable environments for the mosquito vector. Though WNV is endemic, the year-to-year risk is highly variable, thus making it difficult to understand the risk for human spillover events. Abatement districts monitor for infected mosquitoes to help understand these potential risks and to help guide our understanding of the risk posed by these observed infected mosquitoes. Creating optimal monitoring networks will provide more informed decision-making tools for abatement districts and policy makers. Investment in these monitoring networks that capture robust observations on mosquito infection rates will allow for environmentally informed inference systems to help guide decision-making and WNV risk. In turn, enhanced decision-making tools allow for faster response times of more targeted and economical surveillance and mosquito population reduction efforts and the overall reduction of WNV transmission. Here we discuss the data streams, their processing, and specifically three ways to calculate WNV infection rates in mosquitoes.

Key words West Nile virus, Arbovirus, Flavivirus, Zoonosis, Disease forecast modelling, Mosquito-borne disease, Vector-borne disease, Mosquito control

Abbreviations

CoC	City of Chicago
MIR	minimum infection rate
MLE	maximum likelihood estimate
MPN	mosquitoes per night
PPN	pools per night
PPPN	positive pools per night
VI	vector index

Fengwei Bai (ed.), *West Nile Virus: Methods and Protocols,*
Methods in Molecular Biology, vol. 2585, https://doi.org/10.1007/978-1-0716-2760-0_16,

1   Introduction

West Nile virus (WNV) is now endemic to much of the United States, and its range continues to expand as climates warm, creating more habitable environments for the mosquito vector [1, 2]. WNV exhibits considerable seasonal and spatial variability dependent on several factors but is substantially influenced by hydrology and temperature [2, 3]. Additionally, this heterogeneity between and within outbreaks is dependent on spatial scale, where regions of a mosquito abatement district may have a higher incidence of WNV than others, driving up the overall incidence in the district. Understanding this heterogeneity is important for the decision-making process regarding when and where public health interventions, such as adulticiding, should take place [4]. This heterogeneity also makes forecasting WNV outbreaks considerably more complicated when attempted at scales more granular than the district or state level.

Further complicating our understanding of the mosquito infection rate and the decision-making process to treat is the sensitivity and specificity of trap-level mosquito data. First and foremost, mosquito abundance is always only relative. Mosquitoes move, and deploying enough traps to fully census a population is not feasible. Additionally, the relative nature of this abundance can skew measures of WNV prevalence in mosquitoes. For example, consider a scenario where we trap in two locations. At one location, we trap 10 mosquitoes and 1 is positive for WNV, while at the second location, we trap 100 mosquitoes, and 1 is positive for WNV. At first glance, the first location would appear to have a higher infection rate, but upon further consideration, we must ask if this is truly a higher incidence or a product of trap success. Are the ten mosquitoes a representation of the true mosquito abundance in the area or did other factors decrease trapping success?

Furthermore, when using the mosquito infection rate of WNV, we must take into account the relationship between the mosquito trapping data stream and the human case data stream [5]. Due to the nature of public health reporting systems, there are inherent lags between these two data streams, with mosquito infection rate often available much sooner than human case reports [5]. These lags are a product of both public health policy at various jurisdiction levels and delayed reporting due to patient's care-seeking behaviors. Regardless, they must be considered [4–6]. Equally important are the temporal lags that must be considered for the environmental drivers of mosquito biology. Warmer, wetter environments favor mosquito development and WNV amplification, but dryer than usual conditions in a geographic location favor amplification and transmission between the mosquito vector and bird hosts, but these deviations in weather patterns must happen at specific points in the population's development [7, 8]. Their effect on the mosquito infection rate and, therefore, the WNV outbreak potential is not apparent until weeks or months after the event. Enhanced decision-making tools that take these two temporal factors into consideration allow for faster response times, more targeted, economical surveillance and mosquito population reduction efforts, and the overall dampening of a WNV outbreak.

This chapter discusses mosquito surveillance data streams, their processing, and specifically three ways to calculate WNV incidence in mosquitoes for use in better and more informed decision-making tools for abatement districts and policy makers.

1.1 Mosquito Infection Exposure Risk

The purpose of mosquito testing is to determine if the virus is present in an area; however, prevalence rates within the mosquito population for WNV can be very low. Given the low prevalence rates of WNV found in the mosquitoes, it is not economically feasible to test individual mosquitoes to get an estimate of the infection rate for a given area, so statistical approaches are needed to help estimate the infection rate from pooled samples. Thus, pool sampling is conducted, with most abatement districts utilizing pools from 10 to 50 mosquitoes. Here we will discuss three approaches to determine the risk of infected mosquitoes: (1) minimum infection rate (MIR), (2) maximum likelihood estimate (MLE), and (3) vector index (VI).

1.2 Minimum Infection Rate

The MIR is a simple approximation for the prevalence of infected mosquitoes. This calculation is limited, and by definition gives the lower bound for the infected number of mosquitoes that were tested. This calculation assumes that only one mosquito is positive per pool of mosquitoes tested. The formula is as follows:

$$\text{MIR} = \frac{\sum p_i}{\sum m_i n_i}$$

Where n is the number of pools sampled, m is the number of mosquitoes in each pool, and p is 1 if the pool is positive.

1.3 Maximum Likelihood Estimation

The MLE assumes that if a pool tests positive, one or more mosquitoes are WNV positive, whereas a negative result indicates all mosquitoes are WNV negative. This is slightly different than the assumption with the MIR, which assumes only one positive mosquito per pool. Of the two, the MLE is considered the more appropriate estimate of infection rate as the number of mosquitoes tested increases. This calculation takes advantage of changes within pool size and estimates a higher infection level than the MIR because the model does not assume only one mosquito in the pool is infected [9]. Here we can estimate the mosquito infection rate using the MLE assuming a binomial distribution and all pooled samples for the county over a given week. Specifically, the log-likelihood equation for data $x = (x_1, x_2, \ldots x_M)$ is:

$$l(p; x) = l(p) = \sum_{i=0}^{M} x_i \ln\left(1 - (1 - p)^{m_i}\right) \ln\left(1 - p\right) \sum_{i=0}^{M} m_i(n_i - x_i)$$

Where x_i is the number of positive samples for a given pool size, m_i is the distinct number of mosquitoes sampled in a pool, n_i is the number of times the distinct pool size was sampled, and M is the number of distinct pool sizes. The solution to p, the MLE of the equation, is obtained using a numerical solver. Here we use the Newton-Raphson method to iteratively compute successive values of p until convergence. This root-finding algorithm finds where the derivative of the line is equal to zero or the maximum of the point. In certain cases, such as all pools being negative or a sparsity of data due to scale, the solver will not converge. This is often a limitation when dealing with a numerical solver where there is little signal to identify the minimum and converge.

Also, when calculating the prevalence, one should understand how trapping effort plays a role and the impact that the number of mosquitoes tested will have on the robustness of these prevalence estimates. When evaluating at the trap level, it may be appropriate to estimate for the annual infection rate; however, this method may not be appropriate at the weekly time scale because of the limited number of mosquitoes. As spatial scales increase and multiple traps are aggregated to values >300 mosquitoes (Fig. 1), the MLE tends to provide a more robust estimate of the prevalence over space for a given week than the MIR [1].

1.4 Vector Index Calculation

Vector index (VI) is a way to combine both the prevalence of infected mosquitoes with the mosquito population number. Theoretically, given an increase in mosquito population with the same prevalence, one would expect this to be an optimal indicator for the risk of WNV spillover. To calculate the VI at time point i, we use the following equation:

$$\boldsymbol{VI_i = P_i N_i}$$

Where P_i is the prevalence of infected mosquitoes calculated by using the MLE or MIR and N_i is the normalized number of mosquitoes trapped for week i or over a defined area.

2   Materials

Pooled WNV testing results corresponding with trap-level surveillance data.

Please see example data at https://data.cityofchicago.org/widgets/jqe8-8r6s.

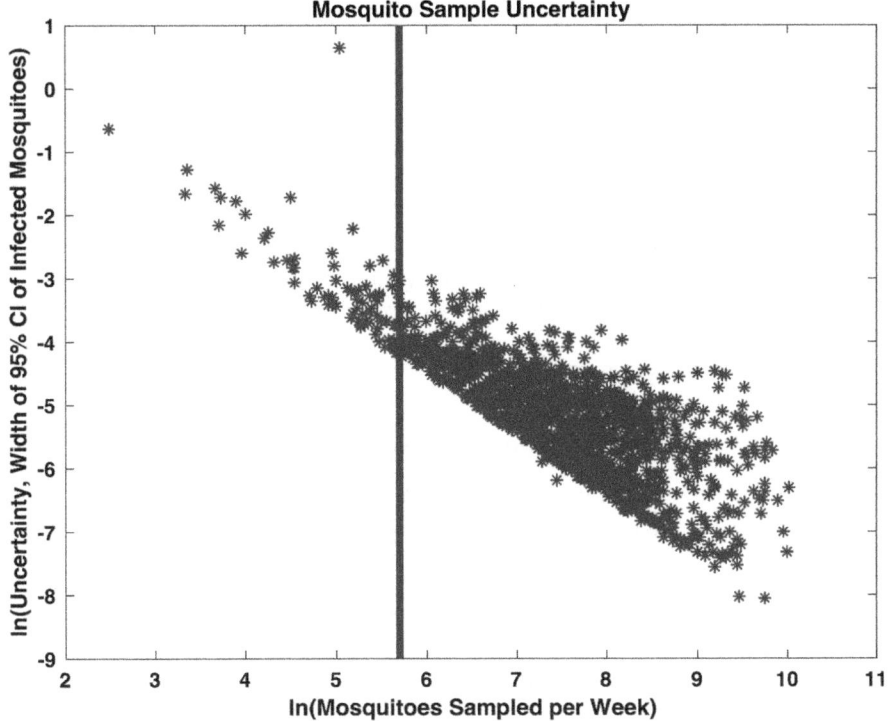

Fig. 1 The natural log of the weekly number of mosquitoes sampled compared to the width of the 95% confidence interval of the estimated proportion of infectious mosquitoes. The uncertainty associated with the proportion of mosquitoes infected becomes sensible when 300 or more mosquitoes are sampled

Mosquito surveillance data in its most basal form begins at the trap level. Traps are set most often at a weekly time scale in areas throughout a district based on when the mosquito season historically begins. The number of traps, their locations, and how often they are set is based on several determining factors.

The number of traps a district chooses to set varies from district to district and is often dependent on a range of determinants including historical precedence, district size, staff capacity, budget, and infection rates. It is often not financially feasible or even sensical to run mosquito traps throughout the year, so traps are typically first deployed in a district based on the historical mosquito season in the region and where WNV has historically been first detected. In very large districts, capacity to deploy traps may be a limiting factor for how many traps are deployed and where, as may be, budget. The smaller the budget and the fewer staff, the fewer the number of traps deployed. Observed mosquito infection rates often impact trapping effort as well. It is not uncommon for districts to increase trapping efforts when WNV is detected in mosquito pools. In addition, location may be affected by access to trapping areas. For example, a district in a more urban setting may have more location constraints related to accessing private property when compared to more rural districts. These factors must be accounted for when using mosquito surveillance data for calculating incidence for use in forecasting WNV. More specifically, trapping effort at the spatial and temporal scales of interest must be taken into account to normalize the data and determine if the data collected is a robust observation of risk. To avoid inflating or deflating the data, the number of positive pools or the number of mosquitoes trapped must be processed to understand the bias in the data, so one can glean the most knowledge from this data. Through the remainder of this chapter, we will provide three statistical techniques and R code for how to process pooled data on a spatial and temporal time frame [R-CITATION].

Additional Considerations when Processing Data

- Mosquito data is available—however infection levels are low—so need tourniquets to help identify the most appropriate value.
- Will trap data produce a false negative?
  - Trap—valid observations or what is the expected error in the observation.
- Which technique for calculating prevalence is most appropriate?

3   Methods

These methods assume a basic literacy of coding in R; for more information and tutorials with R please see, https://www.r-project.org/. Often data sets will come with a number of variables that are not useful or not relevant for determining risk. Additionally, they will have errors and inconsistencies—misspelled enteries, species names, reversed coordinates, etc.

The first step in processing your data will be to evaluate it for such mistakes and either correct or remove them. It is much easier to do this when you first process the data than have to do it retroactively when you realize they are affecting your calculations later in your analysis, or worse never recognize their effect and produce inaccurate results. The City of Chicago (CoC) takes great care in cleaning its data before publishing online, so we will not need to worry about this in the ezample data set.

Next, you will want to evaluate and subset by species capable of transmitting WNV. The primary vector species of concern when discussing WNV are *Culex tarsalis* and *Cx. pipiens/Cx. quinquefasciatus*, but other *Cx.* species have been shown to be capable of carrying WNV. The importance of these species varies by location, but these are the primary vectors of WNV in the United States. Additionally, understanding the local vector populations' feeding and breeding habits can provide insight into WNV amplification and risk of spillover to humans in a specific region. For example, *Cx. tarsalis* generally prefer to lay their egg rafts in clean water often with vegetation such as rice fields, while *Cx. pipiens/Cx. quinquefasciatus* will oviposit in stagnant, biologically rich aquatic environments such as stagnant irrigation ditches and stormwater drainage basins. Understanding these differences and the local ecology will allow you to better understand local transmission dynamics. To do this, determine which species have ever tested positive in your data set. For the CoC data, *Culex pipiens/Culex restuans, Culex restuans, Culex pipiens, Culex territans*, and *Culex salinarius* have historically tested positive. We will then want to subset the data by these species and aggregate it to the spatial and the desired temporal resolution. Note the uncertainty related to the different spatial scale and sampling effort will change the resulting value. We will also need to drop out variables we do not intend to use to calculate risk. In this example, for simplicity, we will use year, week, and trap level, as well as aggregate known competent vector species. We now have a cleaned data set aggregated to our desired temporal and spatial scales.

Please follow the following annotated code for complete step-by-step methods for cleaning and structuring the data as well as calculating the MLE and various measures of mosquito abundance and vector index.

R Code

```
#################################################################-
########
#STATISTICAL TOOLS FOR WEST NILE VIRUS DISEASE ANALYSIS
#EXAMPLE SCRIPT
#WARD, SOREK-HAMER, VEMURI, DEFELICE- 2022
#################################################################-
########
#CITY OF CHICAGO WNV SURVIELLANCE DATA
#https://data.cityofchicago.org/widgets/jqe8-8r6s
#################################################################-
########

#SET WORKING DIRECTORY AND DATE
old.wd <- getwd()
setwd("~YOUR_WORKING_DIRECTORY_HERE")
currentdate <- Sys.Date()

#LOAD PACKGES
suppressPackageStartupMessages({
 library(data.table)
 library(dplyr)
 library(ggplot2)
})
```

```
##############################################################-
########
#STEP ONE
#CLEAN AND FORMAT SURVEILLANCE DATA FOR CALCULATING MLE & MIR
##############################################################-
########

#READ DATA INTO DATA TABLE (OUTCOME VARIABLE SIZE: 32878X13,
YEARS 2002:2021)
dt <- fread("West_Nile_Virus__WNV__Mosquito_Test_Results.
csv")
```

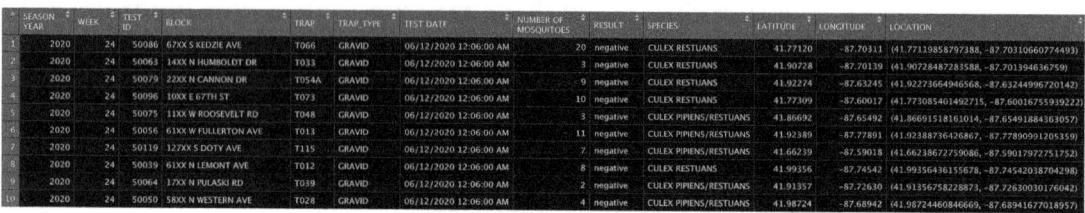

```
#REMOVE OBSERVATIONS WITHOUT COORDINATES (27990X13)
dt <- dt[LONGITUDE != "NA",]

#CREATE DATA TABLE OF TRAP COORDINATES FOR USE LATER (164X3)
coords <- unique(dt[, c("TRAP", "LATITUDE", "LONGITUDE")])

#BRING FORWARD DESIRED VARIABLES AND ADJUST NAMES (27990X7)
sub.dt <- dt[, c("SEASON YEAR", "WEEK", "TRAP", "TEST DATE",
"NUMBER OF MOSQUITOES", "RESULT", "SPECIES")]
sub.dt <- setnames(sub.dt, c("SEASON YEAR", "WEEK", "TRAP",
"TEST DATE", "NUMBER OF MOSQUITOES", "RESULT", "SPECIES"),
 c("year", "week", "trap", "date", "number_of_mosquitoes",
"result", "species"))

#DETERMINE WICH SPECIES HAVE EVER BEEN POSITIVE(2091X7)
temp <- sub.dt[result == "positive"]
unique(temp$species)

#SUBSET BY COMPETENT SPECIES, NOTE; FOR SIMPLICITY
#WE ARE COMBINING COMPETENT SPECIES FOR THIS EXAMPLE (27916X6)
sub.dt <- sub.dt[species == "CULEX PIPIENS/RESTUANS" | species
== "CULEX RESTUANS" | species == "CULEX PIPIENS" |
 species == "CULEX TERRITANS" | species == "CULEX SALINARIUS"]
sub.dt <- sub.dt[, c("date", "year", "week", "trap", "number_-
of_mosquitoes", "result")]
rm(temp)

#FOR THIS EXAMPLE SUBSET THE DATA FOR YEARS 2009 AND 2012 FOR
```

```
ANALYSIS (3936X6)
sub.dt <- sub.dt[year == 2009 | year == 2012]

#REMOVE ANY OBSERVATIONS WITH 'NA' IN THE RESULT (NOTE; THERE
ARE NONE IN THIS DATA BUT IT'S GOOD PRACTICE TO CONFIRM)
sub.dt <- sub.dt[which(sub.dt$result != "NA")]

#ADJUST THE DATE FOR THE SOLVER AND ADD MONTH VARIABLE (3936X6)
sub.dt$date <- as.Date(sub.dt$date, format = "%m/%d/%Y")
sub.dt[, '='(month = month(date))]
sub.dt1 <- sub.dt[, c("year", "month", "week", "trap", "num-
ber_of_mosquitoes", "result")]

#ADD AGENCY VARIABLE (3936X7)
sub.dt1$agency <- "COC"

#SAVE NEW DATA TABLE AS CSV FOR REFERENCE
FileName <- paste("CoC_MLE_Formated_", currentdate, ".csv",
sep = "")
write.csv(sub.dt1, file = FileName, row.names = FALSE)
```

	year	month	week	trap	number_of_mosquitoes	result	agency
1	2009	7	30	T226	16	negative	COC
2	2012	9	39	T048	1	negative	COC
3	2012	8	31	T147	6	negative	COC
4	2012	6	25	T002	11	negative	COC
5	2012	7	27	T011	50	negative	COC
6	2012	7	27	T009	34	positive	COC
7	2009	7	26	T054A	2	negative	COC
8	2009	7	28	T054A	2	negative	COC
9	2009	7	28	T225	4	negative	COC
10	2012	8	32	T079	2	negative	COC

```
###############################################################-
########
#STEP TWO
#CLEAN AND FORMAT SURVEILLANCE DATA FOR USE IN POPULATION
ANALYSIS
###############################################################-
########

#CREATE BINARY RESULTS VARIABLE (3936X8)
sub.dt[result == "negative", positive_pools = 0]
sub.dt[result == "positive", positive_pools = 1]

#DROP UNNEEDED VARAIBLES AND ADD A DUMMY VARIABLE FOR # OF
```

```
POOLS (3936X8)
sub.dt$pools <- 1:1
sub.dt <- sub.dt[, c("year", "week", "trap", "date", "number_-
of_mosquitoes", "month", "positive_pools",
 "pools")]

#SUM ON YEAR, WEEK, DATE AND TRAP AND ADD DUMMY VARIABLE FOR #
OF TRAP WEEKS PER TRAP PER YEAR (1907X9)
sub.dt <- sub.dt[,lapply(.SD,sum), by = c("year", "week",
"trap", "date", "month")]
sub.dt$trap_nights <- 2:2

#BRING FORWARD DESIRED VARIABLES AND SUM BY YEAR AND TRAP
(1897X8)
sub.dt <- sub.dt[, c("year", "week", "trap", "number_of_mos-
quitoes", "month", "positive_pools",
 "pools", "trap_nights")]
sub.dt <- sub.dt[,lapply(.SD,sum), by = c("year", "week",
"trap", "month")]

#CALCULATE NUMBER OF MOSQUITOES PER TRAP NIGHT PER YEAR
(1897X9)
sub.dt$MPN <- sub.dt$number_of_mosquitoes/sub.dt$trap_nights

#CALCULATE NUMBER OF POOLS PER TRAP NIGHT PER YEAR (1897X10)
sub.dt$PPN <- sub.dt$pools/sub.dt$trap_nights

#CALCULATE POSITIVE POOLS PER TRAP NIGHT PER YEAR (1897X11)
sub.dt$PPPN <- sub.dt$positive_pools/sub.dt$trap_nights

#ADD COORDINATES (LAT/LONG FORMAT) TO FORMATTED DATA (1897X13)
coords <- setnames(coords, c("TRAP", "LATITUDE", "LONGITUDE"),
 c("trap", "latitude", "longitude"))
sub.dt <- left_join(sub.dt, coords, by = "trap")

#SAVE NEW DATA TABLE AS CSV FOR REFERENCE
FileName <- paste("CoC_POP_Formated_", currentdate, ".csv",
sep = "")
write.csv(sub.dt, file = FileName, row.names = FALSE)
```

	year	week	trap	month	number_of_mosquitoes	positive_pools	pools	trap_nights	MPN	PPN	PPPN	PI	VI	latitude	longitude
1	2009	30	T226	7	34	0	2	2	17.0	1.00	0.00	0.000000000	0.00	41.79437	-87.64893
2	2012	39	T048	9	9	0	3	2	4.5	1.50	0.00	0.000000000	0.00	41.86692	-87.65492
3	2012	31	T147	8	27	1	3	2	13.5	1.50	0.50	0.037037037	0.50	41.93279	-87.70024
4	2012	25	T002	6	11	0	1	2	5.5	0.50	0.00	0.000000000	0.00	41.95630	-87.79752
5	2012	27	T011	7	444	1	10	2	222.0	5.00	0.50	0.002252252	0.50	41.94596	-87.83294
6	2012	27	T009	7	59	1	2	2	29.5	1.00	0.50	0.016949153	0.50	41.98859	-87.85447
7	2009	26	T054A	7	13	0	2	2	6.5	1.00	0.00	0.000000000	0.00	41.92274	-87.63245
8	2009	28	T054A	7	5	0	2	2	2.5	1.00	0.00	0.000000000	0.00	41.92274	-87.63245
9	2009	28	T225	7	55	0	3	2	27.5	1.50	0.00	0.000000000	0.00	41.74267	-87.73155
10	2012	32	T079	8	2	0	1	2	1.0	0.50	0.00	0.000000000	0.00	41.76575	-87.56247

```
#CLEAR GLOBAL ENVIRONMENT LEAVING LOADED PACKAGES (NOTE; THIS
IS NOT NECESSARY BUT DEPENDING ON YOUR COMPUTER AND THE SIZE OF
YUR DATA SET MAY SAVE COMPUTATION TIME)
rm(list = ls())
gc()


##############################################################-
########
#STEP THREE
#RUN SOLVERS AND CALCULATE MLE & MIR
##############################################################-
########

#LOAD FUNCTIONS AND SOLVERS
#####
#FUNCTIONS TO GENERATE MLW FOR ANY TIME AND SPATIAL SCALE

mle_fun <- function(S){
 if(sum(S$result== "positive") > 0){
  MC=sort(unique(S$number_of_mosquitoes)) # unique mosquito
pools
  n=length(MC); # different baseline number of mosquitoes
sampled
  Data_imPool=matrix(0,nrow=n,ncol=4)

  # subset based on number of mosquitoes in a pool and number of
those pools that are positive
  for (j in 1:n) {
  sn=MC[j]==S$number_of_mosquitoes # number of mosquitoes in a
pool of pool size x
  Ssub<-S[sn,]
  Data_imPool[j,1]=MC[j]; # pool of size x
  Data_imPool[j,2]=sum(sn); # number of pools with that number
of mos
  Data_imPool[j,3]=sum(Ssub$result=="positive")} #how many of
those are positive

   g_mu=sum(Data_imPool[,3])/sum(Data_imPool[,2]*Data_imPool
[,1]) # start an MIR location
```

```
if (sum(Data_imPool[,2]) == sum(Data_imPool[,3])){
MLE_CI=MIR_pool_sampling(guess_mu=g_mu,Data = Data_imPool)
}else {
MLE_CI=MLE_WNVpool_95CI(guess_mu=g_mu,Data = Data_imPool)}
} else {
 MLE_CI=c(0,0,0,0,sum(S$number_of_mosquitoes),0)  # No infec-
tions
 }
 MLE_CI
}


#ADJUST SPATIAL AND TEMPORAL STEPS TO MATCH YOUR DATA
space_time_mle <- function(IM_data,
 temporal_step = c('week','month','year'),
 spatial_step = c('trap','agency')){
 if(isFALSE(is.data.table(IM_data)))setDT(IM_data)
 IM_copy = IM_data

 time_var <- match.arg(temporal_step)
 spat_step <- sort(IM_copy[,unique(get(spatial_step))])
 spat_step <- spat_step[!(spat_step == "NA")]
 time_step <- IM_copy[,unique(get(time_var))]

 spatial_temporal_dt <- IM_copy[,expand.grid(spat_step,time_-
step)]
 setDT(spatial_temporal_dt)
 spatial_temporal_id <- spatial_temporal_dt[,paste(Var1,
Var2,
 sep = "_")]

 IM_copy[,st_id = paste(get(spatial_step),
 get(temporal_step), sep = "_")]

 stID <- intersect(spatial_temporal_id,unique(IM_copy$st_id))

 wki_MLE_calc <- sapply(stID, function(x){
 if(x %in% IM_copy$st_id){
 message(sprintf("Running MLE function for Spatial_temporal_ID
%s",x))
 mle_fun(S = IM_copy[st_id == x])
 }
 }) %>% t()
 wki_MLE_calc=data.table(wki_MLE_calc)
 setnames(wki_MLE_calc,c("MIR","MLE","MLE_L95","MLE_U95","-
number_of_mosquitoes","variance_mle"))
 wki_MLE_calc[,c(substitute(spatial_step),
 substitute(time_var)) = tstrsplit(stID,"_")[]]
 wki_MLE_calc[,substitute(time_var) = as.numeric(get(time_-
```

```
var))]
}

###
MLE_WNVpool_95CI<-function(guess_mu,Data){
 mi=Data[,1];
 ni=Data[,2];
 xi=Data[,3];
 p=Newton_Raphson_solver(Data,guess_mu);

 # find lower
 ps=p/10;
 ph=p;
 pl=p/100;
 Sp=-1/(1-p)*sum(mi*xi/(1-(1-p)^mi)-mi*ni);
 Ip=sum(((mi^2)*ni*(1-p)^(mi-2))/(1-(1-p)^mi));
 zp=Sp/sqrt(Ip);

 # OEV2=Ip^-1;
 while (!(zp> -1.96001 & zp< -1.95999) ){
 Sp=-1/(1-ps)*sum(mi*xi/(1-(1-ps)^mi)-mi*ni);
 Ip=sum(((mi^2)*ni*(1-ps)^(mi-2))/(1-(1-ps)^mi));
 zp=Sp/sqrt(Ip);
 if (zp > -1.96){
 ph=ps
 ps=1/2*(pl+ph)
 pl=pl}

 else if (zp < -1.96){
 pl=ps;
 ps=1/2*(pl+ph);
 ph=ph;}}

 pL95=ps;
 ps=p+ps;
 ph=p+10*ps;
 pl=p;
 Sp=-1/(1-p)*sum(mi*xi/(1-(1-p)^mi)-mi*ni);
 Ip=sum(((mi^2)*ni*(1-p)^(mi-2))/(1-(1-p)^mi));
 zp=Sp/sqrt(Ip);

 count=0;
 while (!(zp< 1.96001 & zp> 1.95999) & count< 1000000){
 count=count+1;
 Sp=-1/(1-ps)*sum(mi*xi/(1-(1-ps)^mi)-mi*ni);
 Ip=sum(((mi^2)*ni*(1-ps)^(mi-2))/(1-(1-ps)^mi));
 zp=Sp/sqrt(Ip);
 if (zp > 1.96){
```

```
ph=ps
ps=1/2*(pl+ph)
pl=pl}

else if (zp < 1.96){
pl=ps;
ps=1/2*(pl+ph);
ph=ph}}

if (count==1000000){
d=p-pL95;
ps=p+d}
pH95=ps;
sN=sqrt(sum(Data[,1]*Data[,2]));
OEV=(((pH95-pL95)*sN)/3.92)^2;
sum_mos=sum(Data[,1]*Data[,2])

MLE95p=c(guess_mu,p,pL95,pH95,sum_mos, OEV)
return(MLE95p)}

#MIR AND WALD INTERVAL
MIR_pool_sampling<-function(guess_mu,Data){
 # CI counducted does not account for the loss in pooling
 sN=sqrt(sum(Data[,1]*Data[,2]))
 MIR_m=guess_mu
 var_mir=guess_mu*(1-guess_mu)/sum(Data[,1]*Data[,2])
 MIR_L95=(guess_mu-1.96*sqrt(var_mir));
 MIR_H95=(guess_mu+1.96*sqrt(var_mir));
 sum_mos=sum(Data[,1]*Data[,2])

 MIR=c(guess_mu,
 MIR_m,
 MIR_L95,
 MIR_H95,
 sum_mos,
 var_mir);

 return(MIR)}

Newton_Raphson_solver<-function(data,x0){
 mi=data[,1];
 ni=data[,2];
 xi=data[,3];
 N=sum(mi*ni);
 # The initial value

 f = function(p) N - sum((mi*xi)/(1-(1-p)^mi)) # The function
whose root we are trying to find
```

```
fprime = function(p) sum((mi^2*xi*(1-p)^(mi-1))/((1-(1-p)^mi)
^2)) # The derivative of f(x)
 tol = 10^(-9) # accuracy desired
 epsilon = 10^(-14) # smallest number
 maxIterations = 1000; # Don't allow the iterations to continue
indefinitely
 haveWeFoundSolution = FALSE # Have not converged to a solution
yet

 for (i in 1:maxIterations){
 y = f(x0)
 yprime = fprime(x0)
 if(abs(yprime) < epsilon) { # Don't want to divide by too
small of a number
 # denominator is too small
 break} # Leave the loop
 x1 = x0 - y/yprime ; # Do Newton's computation
 if(abs(x1 - x0)/abs(x1) < tol){ # If the result is within the
desired tolerance
 haveWeFoundSolution = TRUE;
 break() } # Done, so leave the loop

 x0 = x1} # Update x0 to start the process again
 #if (haveWeFoundSolution) {print('x1 is a solution within
tolerance and maximum number of iterations')}
 if (!haveWeFoundSolution) {print('did not converge')}
 return(x1)
}
 #####

#####
#CALCULATE MLE
#####
old.wd <- getwd()
setwd("~YOURWORKINGDIRECTORY")
currentdate <- Sys.Date()

#READ DATA INTO DATA TABLE. NOTE; ENSURE CORRECT VERSION IS
LOADED BY DATE (3936X7)
MLE <- fread("CoC_MLE_Formated_DATE.csv")
MLE <- na.omit(MLE) #NOTE; THERE ARE NONE IN THIS DATA BUT GOOD
PRECTIVE TO CONFIRM

years <- MLE[,unique(MLE$year)]
setattr(years, 'names', years)

#ANNUAL MLE AT THE AGENCY LEVEL (2X9); NOTE FUNCTIONS WILL SHOW
```

```
PROGRESS IN
# CONSUL AS THEY RUN
COC_Agency_MLE_output_yr <- rbindlist(lapply(years, function
(i.year){
 message(sprintf("Running MLE function for year %s",i.year))
  IM_data = MLE[year == i.year]
  mle_out <- space_time_mle(IM_data, spatial_step = 'agency',
temporal_step = 'year')
}),idcol = "year")

FileName <- paste("Annual_MLE_Agency_", currentdate, ".csv",
sep = "")
write.csv(COC_Agency_MLE_output_yr, file = FileName, row.names
= FALSE)
```

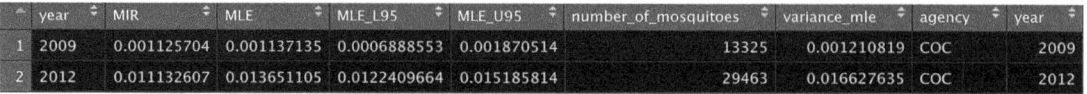

	year	MIR	MLE	MLE_L95	MLE_U95	number_of_mosquitoes	variance_mle	agency	year
1	2009	0.001125704	0.001137135	0.0006888553	0.001870514	13325	0.001210819	COC	2009
2	2012	0.011132607	0.013651105	0.0122409664	0.015185814	29463	0.016627635	COC	2012

```
#WEEKLY AT THE AGENCY LEVEL (37X9); NOTE FUNCTIONS WILL SHOW
PROGRESS IN
# CONSUL AS THEY RUN
COC_Agency_MLE_output_wk <- rbindlist(lapply(years, function
(i.year){
 message(sprintf("Running MLE function for year %s",i.year))
  IM_data = MLE[year == i.year]
  mle_out <- space_time_mle(IM_data, spatial_step = 'agency',
temporal_step = 'week')
}),idcol = "year")

FileName <- paste("Weekly_MLE_Agency_", currentdate, ".csv",
sep = "")
write.csv(COC_Agency_MLE_output_wk, file = FileName, row.names
= FALSE)
```

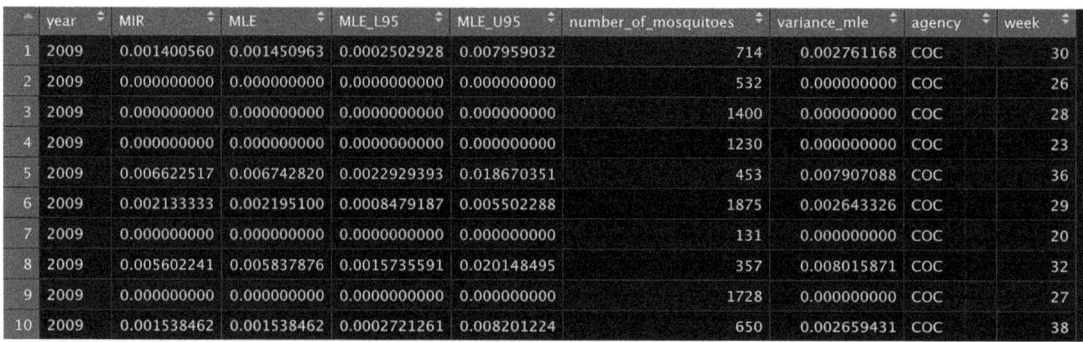

	year	MIR	MLE	MLE_L95	MLE_U95	number_of_mosquitoes	variance_mle	agency	week
1	2009	0.001400560	0.001450963	0.0002502928	0.007959032	714	0.002761168	COC	30
2	2009	0.000000000	0.000000000	0.0000000000	0.000000000	532	0.000000000	COC	26
3	2009	0.000000000	0.000000000	0.0000000000	0.000000000	1400	0.000000000	COC	28
4	2009	0.000000000	0.000000000	0.0000000000	0.000000000	1230	0.000000000	COC	23
5	2009	0.006622517	0.006742820	0.0022929393	0.018670351	453	0.007907088	COC	36
6	2009	0.002133333	0.002195100	0.0008479187	0.005502288	1875	0.002643326	COC	29
7	2009	0.000000000	0.000000000	0.0000000000	0.000000000	131	0.000000000	COC	20
8	2009	0.005602241	0.005837876	0.0015735591	0.020148495	357	0.008015871	COC	32
9	2009	0.000000000	0.000000000	0.0000000000	0.000000000	1728	0.000000000	COC	27
10	2009	0.001538462	0.001538462	0.0002721261	0.008201224	650	0.002659431	COC	38

```
#############################################################-
```

```
########
#STEP FOUR
#CALCULATE VI AT DESIRED SCALES
############################################################-
########

#READ POPULATION DATA INTO DATA TABLE. NOTE; ENSURE CORRECT
VERSION IS LOADED BY DATE (1897X13)
POP <- fread("CoC_POP_Formated_2022-03-16.csv")
POP <- na.omit(POP) #NOTE; THERE ARE NONE IN THIS DATA BUT GOOD
PRECTIVE TO CONFIRM

#BRING FORWARD DESIRED VARIABLES AND AVERAGE BY YEAR (2X2)
POP_YR <- POP[, c("year", "number_of_mosquitoes")]
POP_YR <- POP_YR[,lapply(.SD,mean), by = c("year")]
POP_YR <- setnames(POP_YR, c("year", "number_of_mosquitoes"),
 c("year", "average_mosquitoes"))

#BRING FORWARD DESIRED VARIABLES AND AVERAGE BY YEAR AND WEEK
(37x3)
POP_WK <- POP[, c("year", "week", "number_of_mosquitoes")]
POP_WK <- POP_WK[,lapply(.SD,mean), by = c("year", "week")]
POP_WK <- setnames(POP_WK, c("year", "week", "number_of_mos-
quitoes"),
 c("year", "week", "average_mosquitoes"))

#MERGE DATA TABLES OF CORRESSPONDING SCALES 7 CALCULATE VI
#ANNUAL (2X10)
ANNUAL_MLE <- COC_Agency_MLE_output_yr[, c("year", "MIR",
"MLE", "MLE_L95", "MLE_U95", "number_of_mosquitoes", "varian-
ce_mle", "agency")]
POP_YR$year <- as.character(POP_YR$year)
ANNUAL <- left_join(ANNUAL_MLE, POP_YR, by = "year")
ANNUAL$VI <- ANNUAL$MIR*ANNUAL$average_mosquitoes

FileName <- paste("COC_ANNUAL_", currentdate, ".csv", sep =
"")
write.csv(ANNUAL, file = FileName, row.names = FALSE)

#WEEKLY (37X11)
WEEK_MLE <- COC_Agency_MLE_output_wk[, c("year", "week",
"MIR", "MLE", "MLE_L95", "MLE_U95", "number_of_mosquitoes",
"variance_mle", "agency")]
POP_WK$year <- as.character(POP_WK$year)
WEEKLY <- left_join(WEEK_MLE, POP_WK, by = c("year", "week"))
WEEKLY$VI <- WEEKLY$MIR*WEEKLY$average_mosquitoes

FileName <- paste("COC_WEEKLY_", currentdate, ".csv", sep =
```

```
"")
write.csv(WEEKLY, file = FileName, row.names = FALSE)

#CLEAR GLOBAL ENVIRONMENT LEAVING LOADED PACKAGES
rm(list = ls())

################################################################-
########
#STEP FIVE
#USEFUL FIGURES
################################################################-
########

#READ IN DATA; FOR THIS EXAMPLE WEEKLY DATA WE JUST CALCULATED
WEEKLY <- fread("COC_WEEKLY_2022-03-16.csv")

#MLE WEEKLY BY YEAR
ggplot() +
 geom_line(data = WEEKLY,
 aes(x = week, y = MLE*1000)) +
  scale_y_continuous(name  =  expression(MLE~(I[M]~"/
"~"1,000"~tested))) +
 facet_wrap(vars(year)) +
 theme_classic() +
 theme(text = element_text(size = 20)) +
 ggsave("weekly_MLE.png",
 width = 12,
 height = 10,
 units = "in",
 device = "png")
```

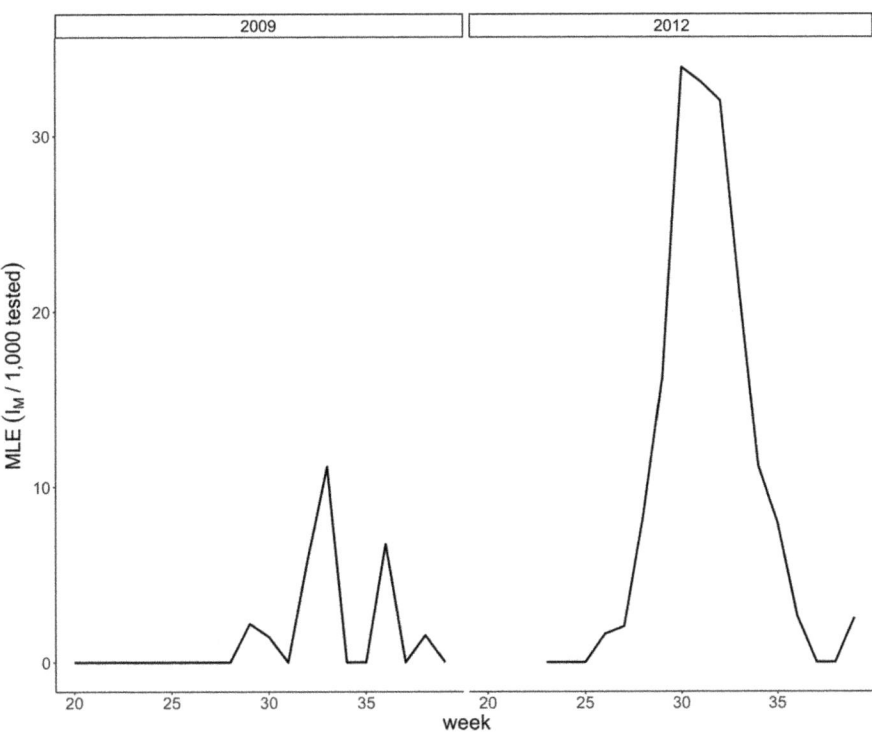

```
#MIR WEEKLY BY YEAR
ggplot() +
 geom_line(data = WEEKLY,
 aes(x = week, y = MIR)) +
 facet_wrap(vars(year)) +
 theme_classic() +
 theme(text = element_text(size = 20)) +
 ggsave("weekly_MIR.png",
 width = 12,
 height = 10,
 units = "in",
 device = "png")
```

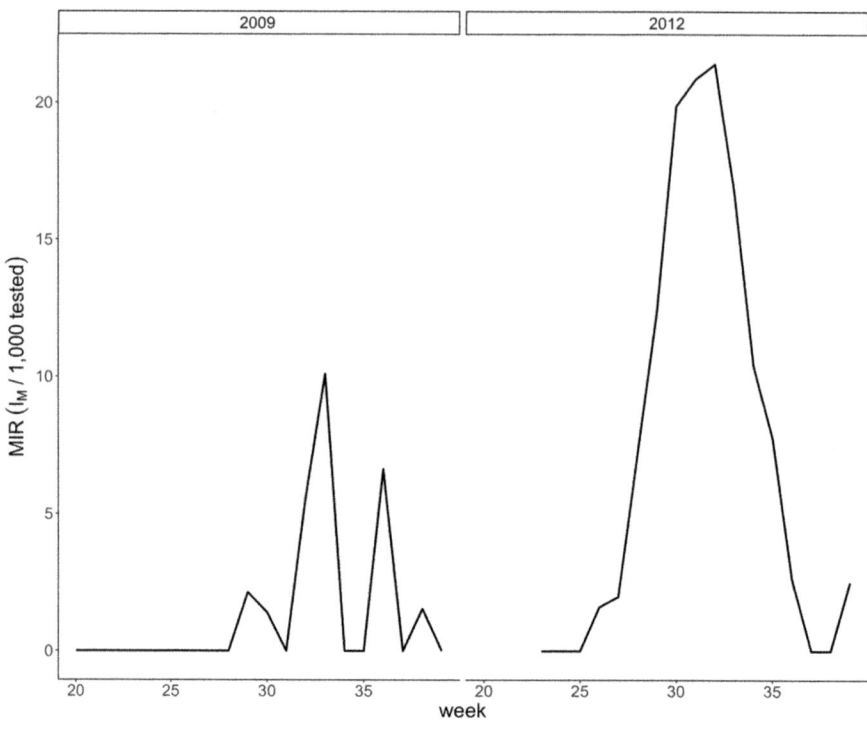

```
#VI WEEKLY BY YEAR
ggplot() +
 geom_line(data = WEEKLY,
 aes(x = week, y = VI)) +
 facet_wrap(vars(year)) +
 theme_classic() +
 theme(text = element_text(size = 20)) +
 ggsave("weekly_VI.png",
 width = 12,
 height = 10,
 units = "in",
 device = "png")
```

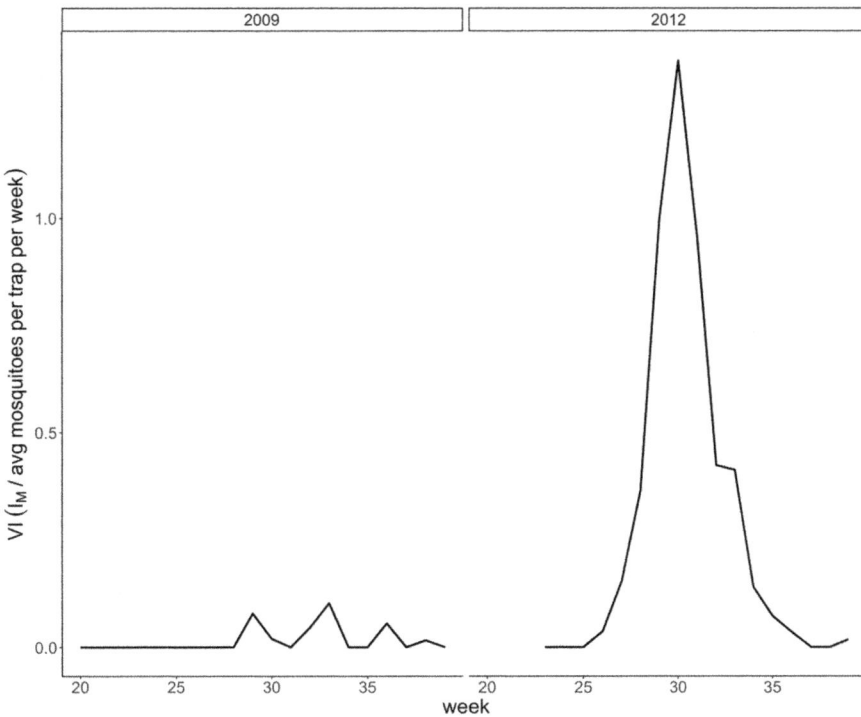

References

1. DeFelice NB, Schneider Z, Little E, Barker C, Caillouet KA, Campbell SR, Damian D, Irwin P, Jones HMP, Townsend J, Shaman J (2018) Use of temperature to improve West Nile virus forecasts. PLoS Comput Biol 14(3). https://doi.org/10.1371/journal.pcbi.1006047

2. Shocket MS, Verwillow AB, Numazu MG, Slamani H, Cohen JM, El Moustaid F et al (2020) Transmission of West Nile and five other temperate mosquito-borne viruses peaks at temperatures between 23 C and 26 C. elife 9: e58511

3. Little E, Campbell SR, Shaman J (2016) Development and validation of a climate-based ensemble prediction model for West Nile virus infection rates in Culex mosquitoes, Suffolk County, New York. Parasit Vectors 9(1):443

4. Barber LM, Schleier JJ, Peterson RK (2010) Economic cost analysis of West Nile virus outbreak, Sacramento county, California, USA, 2005. Emerg Infect Dis 16(3):480–486

5. DeFelice NB, Birger R, DeFelice N, Gagner A, Campbell SR, Romano C, Santoriello M, Henke J, Wittie J, Cole B, Kaiser C, Shaman J (2019) Modeling and surveillance of reporting delays of mosquitoes and humans infected with West Nile virus and associations with accuracy of West Nile virus forecasts. JAMA Netw Open 2(4):e193175. https://doi.org/10.1001/jamanetworkopen.2019.3175

6. Nasci R, Fischer M, Lindsey N, Lanciotti R, Savage H, Komar N et al (2013) West Nile virus in the United States: guidelines for surveillance, prevention, and control. Centers for Disease Control and Prevention, Atlanta

7. Paull SH, Horton DE, Ashfaq M, Rastogi D, Kramer LD, Diffenbaugh NS et al (2017) Drought and immunity determine the intensity of West Nile virus epidemics and climate change impacts. Proc R Soc B 284(1848):20162078

8. Shaman J, Day JF, Stieglitz M (2005) Drought-induced amplification and epidemic transmission of West Nile virus in southern Florida. J Med Entomol 42(2):134–141

9. Gu W, Lampman R, Novak R (2004) Assessment of arbovirus vector infection rates using variable size pooling. Med Vet Entomol 18(2): 200–204

Analytical Approaches to Uncover Genetic Associations for Rare Outcomes: Lessons from West Nile Neuroinvasive Disease

Megan E. Cahill and Ruth R. Montgomery

Abstract

West Nile viral infection causes severe neuroinvasive disease in less than 1% of infected humans. There are no targeted therapeutics for this serious and potentially fatal disease, and to date no vaccine has been approved for humans. With climate change expected to result in rising incidence of West Nile and other related vector-borne viral infections, there is an increasing need to identify those at risk for serious disease and potential leads for therapeutic and vaccine development. Genetic variation, particularly in genes whose products are either directly or indirectly connected to immune response to infections, is a critical avenue of investigation to identify those at higher risk of clinically apparent West Nile infection. Given the small percent of infections that progress to severe disease and the relatively low numbers of reported infections, it is challenging to conduct well-powered studies to identify genetic factors associated with more severe outcomes. In this chapter, we outline several approaches with the objective to take full advantage of all available data in order to identify genetic factors which lead to increased risk of severe West Nile neuroinvasive disease. These methods are generalizable to other conditions with limited cohort size and rare outcomes.

Key words West Nile virus, West Nile neuroinvasive disease, Candidate gene study, Genome-wide association study, Imputation, Gene-gene interactions, Population controls

1 Introduction

Humans infected with West Nile virus (WNV) may be asymptomatic or develop mild or severe disease. West Nile neuroinvasive disease (WNND) can be characterized as meningitis, encephalitis, or acute flaccid paralysis and is estimated to occur in less than 1% of WNV-infected individuals [1, 2]. Given that such a small proportion of WNV-infected individuals develop severe disease, genetic factors are likely to play a role in disease pathogenesis, with some genetic variants leading to greater risk of severe disease or in other cases to a reduced risk. With no approved vaccines or disease-

Fengwei Bai (ed.), *West Nile Virus: Methods and Protocols*,
Methods in Molecular Biology, vol. 2585, https://doi.org/10.1007/978-1-0716-2760-0_17,

specific therapeutics, it is important to understand why some individuals are at increased risk for severe, potentially fatal disease in order to guide development of much-needed vaccines and therapeutics.

To assess genetic predisposition to severe WNND, the most common approaches are case-control studies examining genetic variants, either in specific genes or genome-wide [3]. In the candidate gene studies, analysis is limited to genes suspected to be involved in an aspect of the disease pathogenesis based on a priori knowledge, such as from laboratory experiments or hypotheses from genetic associations identified in similar diseases. Genome-wide association studies (GWAS) are a widely used genetic tool to analyze all genotyped variants for an association with the phenotype of interest. For example, recent GWAS studies on a large cohort of subjects have identified genetic predisposition to severe COVID-19 infection including 2′-5′-oligoadenylate synthetase 1 (OAS1) and tyrosine kinase 2 (TYK2) genes [4, 5]. Importantly, GWAS generally include 1000s of subjects and have a stringent p-value threshold (usually $p < 5 \times 10^{-8}$) to compensate for the large number of statistical analyses. For less common conditions with smaller cohorts of subjects, such as WNV, initial GWAS analysis may be underpowered for definitive identification, and thus additional methods may be needed to augment detection of relevant factors.

To date, only one GWAS has been conducted to analyze associations with severe WNND [6]. Set in a North American sample including WNND cases ($n = 560$ cases) and WNV-positive, mildly ill individuals as the controls ($n = 950$ controls), this study identified single nucleotide polymorphisms (SNPs) within the replication factor C1 (RFC1), sodium channel neuronal type I α subunit (SCN1A), and ananyl aminopeptidase (ANPEP) genes which may be associated with WNND. At least 12 candidate genes studies have also been conducted, focused on immune-related genes such as C-C chemokine receptor type 5 (CCR5) and interferon regulatory transcription factor 3 (IRF3) [3]. Several of these gene-specific studies have been conducted in Greece, Israel, and Macedonia, and while they include diverse subject demographics, most of the sample sizes were below 500. These studies provide evidence of potential genetic associations with WNND, but additional research is needed to confirm the associations and to describe further the genetic architecture of this disease.

There are unique hurdles for analyzing genetic associations with WNND or other small disease cohorts. Most notably, the potential pool of cases is severely limited given a fairly low incidence of disease and a small fraction of infected individuals which progress to WNND. Studies conducted to date have had small sample sizes and thus limited power to identify associations. Additionally, the majority of studies conducted thus far have been in North

American populations of European descent, despite the fact that the virus has been isolated in every continent except Antarctica [7]. With the widespread distribution of WNV and an expected increase in transmission due to climate change [8], it is critical both to conduct new studies and to extract as much information as possible from previously conducted studies. Several innovations in analytic methods can be applied to existing GWAS datasets to elucidate further the role of genetic factors in WNND pathogenesis and to identify novel associations to this potentially fatal disease. Here we outline methodological approaches for in-depth analysis of existing GWAS data.

2   Materials

The only required material is directly genotyped data for a study of WNND (or similarly rare outcome) cases and controls. Depending on the approach(es) chosen, many software programs are freely available and noted below at the corresponding step.

3   Methods

To conduct a GWAS analysis requires sample collection from cases and appropriate controls, DNA sequencing, quality control steps, and initial univariate analysis [9]. For a rare cohort, where statistical power may be lower due to limited sample size, further analyses following the initial univariate analysis of the dataset can identify novel variants associated with severe WNND or other similarly rare outcomes (*see* **Note 1**). Depending on the available data and study aim, researchers may choose to use imputation (Subheading 3.2), analyze gene-gene interactions (Subheading 3.3), and/or incorporate population controls (Subheading 3.4) to identify novel associations (Fig. 1).

3.1   Choosing a Validated GWAS Dataset

The approaches outlined below should be conducted on existing cleaned genotype data, either directly prepared by study researchers or requested through data repositories (*see* Subheading 3.4.1). GWAS datasets contain sensitive personal information and are often large files, and a secure, robust computing environment that meets laws and institutional policies is required. Prior to analysis, researchers should clean the dataset through quality control (QC) steps, such as application of thresholds for individual and variant call rates [10]. For WNV, we worked with the sole WNV GWAS dataset and used a high-performance computing cluster with a Linux operating system to execute the below analyses. QC steps ensured analyses were on SNPs and individuals with <5% missing data, and we excluded any SNPs that deviated significantly from Hardy Weinberg Equilibrium [11].

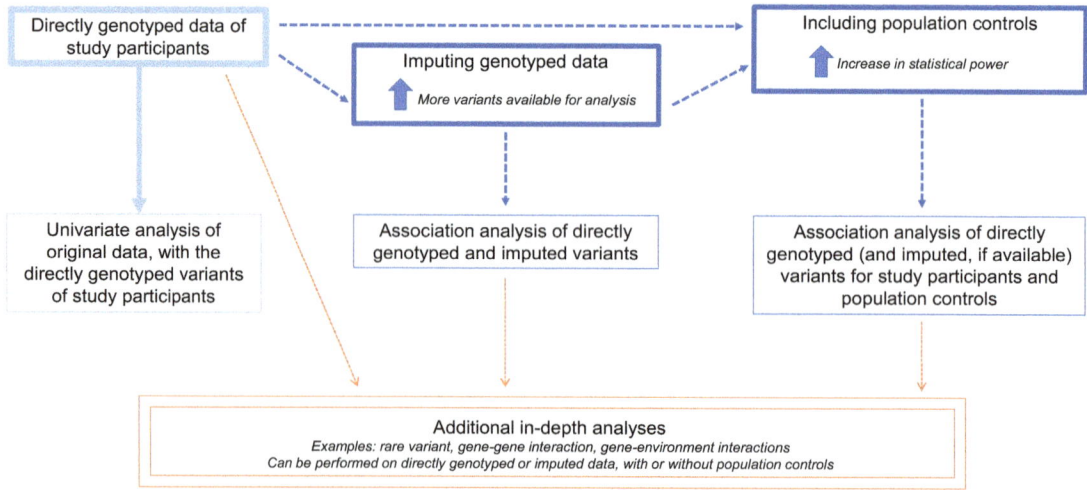

Fig. 1 Overview of analytical approaches to uncover genetic associations for rare outcomes. Beginning with directly genotyped data of a study sample, researchers can use imputation to increase the number of variants available for analysis and/or population controls to increase the statistical power. In-depth analysis may also include rare variant, gene-gene interaction, and gene-environment interactions on these datasets

3.2 Imputation

Imputation estimates variants that were not genotyped in the original study based on comparison to larger reference datasets, which consist of directly genotyped samples. Imputation increases the ability to detect novel loci for an association with WNND or another phenotype of interest. With imputation, the number of common variants available for analysis may increase by eightfold or greater [11, 12]. The steps below outline a hands-on and versatile approach for imputing data. Researchers with limited coding experience may choose to use the Michigan Imputation Server, which provides a free and user-friendly online platform to impute study data [13, 14].

3.2.1 Build Conversion and Pre-phasing

1. If the samples were collected or processed using older technology, the data may need to be converted to a more recent build for imputation and analysis (*see* **Note 2**). The University of California, Santa Cruz, provides the Batch Coordinate Conversion (LiftOver) tool to quickly and easily convert data [15, 16].

2. Pre-phasing, or estimating haplotypes, improves the accuracy and efficiency of imputation. ShapeIt is a popular tool for pre-phasing and provides output that can then be used in Impute2 for the imputation step [17–20].

3.2.2 Imputation Using Software Such as Impute2

Additional software such as Beagle and Minimac are well-used [21, 22], with comparable accuracy depending on the ancestry of the study participants [23, 24].

1. Imputation relies on a reference panel to estimate ungenotyped variants, and the 1000 Genomes is a validated and well-established option [25–27]. Reference panels should match the ancestry of the study participants as closely as possible for accurate imputation [28].

2. Impute2 recommends imputing chunks of chromosomes to increase computational efficiency and imputation accuracy [29]. One strategy is to divide the chromosomes in equal-sized chunks; however, some chunks may have too few SNPs for accurate analysis and/or crossover the centromere. Research groups from the University of Michigan Center for Statistical Genetics and University of California, Santa Cruz Genome Browser have provided detailed instructions on chromosome chunk intervals and centromere locations, respectively [30, 31].

3. Impute2 assesses imputed quality for each variant with an INFO score, ranging between 0 and 1.0, with a higher score indicating increased certainty [32]. To ensure only variants imputed with high certainty are included in the analysis, a cutoff threshold such as 0.7 should be applied to the INFO scores.

3.2.3 Univariate Analysis in PLINK

The directly genotyped and high-quality imputed variants can be analyzed for an association with WNND in PLINK or a similar software, with adjustments as needed for any covariates of interest [33].

3.3 Gene-Gene Interactions

Another valuable tool to identify genetic influences in a rare disease or small cohort is assessing gene-gene interactions, also called epistasis. This method adds to the analysis by identifying pairs of genes which together may play a role in the outcome under study [34]. Identifying the combined effects of variation at two loci of interest can provide additional insight into the genetic architecture of WNND.

3.3.1 Minor Allele Frequency (MAF) Cutoff

To assess gene-gene interactions, the dataset should only include common variants with minor allele frequency (MAF) above a threshold such as 10%. MAF is the estimated frequency of the second most common allele for a variant in a population of matching ancestry; more common variants have a higher MAF, while rare variants have low MAF values [35]. For this analysis, only common variants with MAF above a threshold such as 10% should be included. Limiting the dataset to variants $\geq 10\%$ MAF can be done using the --maf flag in PLINK [33].

3.3.2 Fast-Epistasis Using Genotype Count Tables in PLINK	1. As an initial step to filter the variants, PLINK's fast-epistasis function can quickly test interactions by comparing genotype count tables between all SNPs with MAF $\geq 10\%$.
	2. Variants with an interaction p-value below a cutoff (such as 5×10^{-8}) can then be included in further, more precise analysis of interactions.

3.3.3 Regular Epistasis Function in PLINK

The regular epistasis function, which employs a logistic regression approach, can then be conducted between SNPs that met the threshold from the fast-epistasis step. Alternatively, interactions can be assessed between one SNP that met the fast-epistasis cutoff and all other genotyped variants.

3.4 Population Controls

Including population controls is another relevant option to enhance the analysis of the role of genetic variation on rare outcomes. The addition of relevant genotype datasets to serve as population controls may increase the study size and statistical power to detect associations. Given the low incidence of WNND, many studies are limited in size and underpowered to examine genome-wide associations. An appropriate population control dataset can be used to increase a study's power to identity potential variants associated with WNND.

3.4.1 Selection of Population Controls

Population controls should closely match the study's population in ancestry, as well as any additional limitations such as age, location, and comorbidities. Possible datasets can be searched through databases like dbGaP [36, 37], which requires an application detailing rationale and proposed use for access to individual-level data.

3.4.2 Matching Quality Control Measures

Data from the population controls should be processed similarly to data from the cases and controls of the original study. Both datasets should have the same process to assess and apply thresholds for missing data at the individual and variant level, minor allele frequency, and deviation from Hardy-Weinberg equilibrium. Both datasets need to be of the same build, with software like Batch Coordinate Conversion (LiftOver) used to update build as needed.

3.4.3 Principal Components and Matching Minor Allele Frequency

1. To adjust for heterogeneity and differences in ancestry of the population controls and study sample, principal components can be calculated and outliers removed using software such as Eigenstrat (*see* **Note 3**) [38, 39].

2. Principal components should be plotted (e.g., principal component 1 by principal component 2 for each individual, using distinct colors to differentiate population controls from study cases and controls) to visually assess similarity of the datasets.

3. To further reduce heterogeneity between the samples, analysis should be limited to variants with MAF within 5% between the population controls and the study controls.

3.4.4 Univariate Analysis   Analysis of the association of variants with WNND can be conducted in PLINK or similar software. Adjustments should be made for principal components and study membership (either population control or WNND study), in addition to any other covariates of interest (such as genetic sex and age).

3.5 Discussion   WNV can cause severe, fatal illness in some infected individuals, with no approved vaccines or targeted therapeutics. Given the difficulty of identifying sufficient cases for a well-powered GWAS, it is necessary to consider novel approaches both for future studies and in order to extract further information to fully utilize previously conducted studies. There are many challenges of working with existing datasets, as some data such as the type of neuroinvasive disease or covariates may be unavailable and older studies may have utilized now-discontinued chips or processing methods. With rapid improvement in technology used in genetic research, there may be a need to update the datasets in order to work with current software, but methods such as the ones we outlined above make this feasible. Here we have identified additional approaches, such as analysis of rare variants or pathways for association with severe disease [40, 41], that can also be utilized on either novel or existing datasets to strengthen the ability to detect genetic variants relevant to the disease cohort (*see* **Note 4**). Gene-environment interactions may also be of value [42, 43], but use may be limited for datasets from previously conducted studies depending on the covariates collected at the time.

When conducting future studies, an emphasis should be placed on greater diversity of study samples to match the global spread of this virus. WNV has been isolated worldwide, and an increasing number of outbreaks have occurred in recent years in Europe and the Middle East [7, 44]. There has been a movement to expand the diversity of GWAS beyond the initial disproportionate concentration on samples of individuals with European-descent [45–47], and there are many tools and reference datasets now available to accurately analyze non-European and admixed populations [26, 48]. With better-powered and more diverse studies, researchers will be better equipped to identify host factors that place individuals at increased risk for severe disease.

While WNV has been isolated worldwide and incidence is expected to continue to rise due to climate change, the number of affected individuals remain small compared to some other diseases. With an estimated 80% of infected individuals remaining asymptomatic [2], there are challenges to identifying and enrolling a sufficient number of cases for a well-powered GWAS. Creative approaches, such as testing the blood of individuals enrolled in other studies for antibodies to WNV or collaborating with blood banks to find current asymptomatic infections [49, 50], can be used to increase sample sizes and to provide asymptomatic controls.

Genetic studies have great potential to guide the development of vaccines and targeted therapeutics, as well as potentially identifying individuals at elevated risk for severe neuroinvasive disease. Continued research through novel analyses of existing datasets using approaches like the ones we outlined above, as well as conducting new, well-powered, and diverse studies, are expected to provide further evidence of genetic factors associated with severe outcomes and novel avenues of investigation for developing methods to prevent and treat WNND. It is challenging to research the role of genetic variation in the development of WNND and other rare conditions given the obstacles noted above, such as case identification and statistical power. Through the use of innovative technology and pooling of samples and resources, significant progress continues to be made toward identifying those at risk and developing preventative and therapeutic measures to protect populations against WNND and other public health threats [51–54].

4 Notes

1. *Use the most recent approaches.* Methodologic and analytical approaches will depend on the existing knowledge at the time of the study, the research question (e.g., whether whole genome or focused on specific genes), and the available study population. Genomic technology advances rapidly, and it is recommended that researchers review recent publications for new approaches prior to designing a study.

2. *Navigate reference builds and program versions.* If using an older dataset, be sure to check that it is in the right version and format for a software, as discussed under Subheading 3.2.1. Older datasets may have been sequenced on a now-outdated reference build. Different software programs will have different formatting requirements; check the program's website for the latest versions.

3. *Address population stratification.* Cases and controls need to be as similar as possible for all factors other than the disease of interest; population stratification will lead to inaccurate or unusable results. The careful selection of controls—whether within the study or as supplemental population controls—is critical. Principal component analysis, outlined in Subheading 3.4.3 above and more extensively in other resources [10, 55, 56], is an important step to identifying and adjusting heterogeneity.

4. *Compare data from closely related cohorts.* Similar phenotypes can be analyzed together for greater insight, especially for rare and related outcomes. Like WNV, dengue is a mosquito-borne flavivirus that causes severe disease in some infected individuals;

we analyzed dengue and WNV GWAS datasets together to identify variants associated with both outcomes. This approach has also been used for more common outcomes, particularly psychiatric conditions [3, 57, 58].

Acknowledgments

This work was supported in part by the US National Institutes of Health (NIH)/National Institute of Allergy and Infectious Diseases (NIAID) Human Immunology Project Consortium (HIPC) award U19 AI 089992. The authors are grateful to Dr. Andrew Dewan for expert guidance and Ms. Xiaomei Wang for valuable support.

References

1. Debiasi RL (2011) West Nile virus neuroinvasive disease. Curr Infect Dis Rep 13(4): 350–359. https://doi.org/10.1007/s11908-011-0193-9

2. Centers for Disease Prevention and Control: West Nile virus: symptoms, diagnosis, & treatment (2018) https://www.cdc.gov/westnile/symptoms/index.html. Accessed December 16, 2021

3. Cahill ME, Conley S, DeWan AT et al (2018) Identification of genetic variants associated with dengue or West Nile virus disease: a systematic review and meta-analysis. BMC Infect Dis 18(1):282. https://doi.org/10.1186/s12879-018-3186-6

4. COVID-19 Host Genetics Initiative (2021) Mapping the human genetic architecture of COVID-19. Nature 600(7889):472–477. https://doi.org/10.1038/s41586-021-03767-x

5. Pairo-Castineira E, Clohisey S, Klaric L et al (2021) Genetic mechanisms of critical illness in COVID-19. Nature 591(7848):92–98. https://doi.org/10.1038/s41586-020-03065-y

6. Loeb M, Eskandarian S, Rupp M et al (2011) Genetic variants and susceptibility to neurological complications following West Nile virus infection. J Infect Dis 204(7):1031–1037. https://doi.org/10.1093/infdis/jir493

7. Chancey C, Grinev A, Volkova E et al (2015) The global ecology and epidemiology of West Nile virus. Biomed Res Int 2015:376230. https://doi.org/10.1155/2015/376230

8. Paz S (2015) Climate change impacts on West Nile virus transmission in a global context. Philos Trans R Soc Lond B Biol Sci 370(1665). https://doi.org/10.1098/rstb.2013.0561

9. Marees AT, de Kluiver H, Stringer S et al (2018) A tutorial on conducting genome-wide association studies: quality control and statistical analysis. Int J Methods Psychiatr Res 27(2):e1608. https://doi.org/10.1002/mpr.1608

10. Turner S, Armstrong LL, Bradford Y, et al (2011) Quality control procedures for genome-wide association studies. Curr Protoc Hum Genet;Chapter 1:Unit1 19. https://doi.org/10.1002/0471142905.hg0119s68

11. Cahill ME, Loeb M, Dewan AT et al (2020) In-depth analysis of genetic variation associated with severe West Nile viral disease. Vaccines (Basel) 8(4). https://doi.org/10.3390/vaccines8040744

12. Wood AR, Perry JR, Tanaka T et al (2013) Imputation of variants from the 1000 genomes project modestly improves known associations and can identify low-frequency variant-phenotype associations undetected by HapMap based imputation. PLoS One 8(5):e64343. https://doi.org/10.1371/journal.pone.0064343

13. Das S, Forer L, Schonherr S et al (2016) Next-generation genotype imputation service and methods. Nat Genet 48(10):1284–1287. https://doi.org/10.1038/ng.3656

14. Christian Fuchsberger LF, Schoenherr S, Das S, Abecasis G (2021) Michigan imputation server: Free next-generation genotype imputation service. https://imputationserver.sph.umich.edu/. Accessed October 2, 2021

15. Kent WJ, Sugnet CW, Furey TS et al (2002) The human genome browser at UCSC.

Genome Res 12(6):996–1006. https://doi.org/10.1101/gr.229102

16. University of California Santa Cruz Genomics Institute: lift genome annotations. https://genome.ucsc.edu/cgi-bin/hgLiftOver. Accessed October 5, 2021

17. Delaneau O, Marchini J, Zagury JF (2011) A linear complexity phasing method for thousands of genomes. Nat Methods 9(2): 179–181. https://doi.org/10.1038/nmeth.1785

18. Delaneau O, Zagury JF, Marchini J (2013) Improved whole-chromosome phasing for disease and population genetic studies. Nat Methods 10(1):5–6. https://doi.org/10.1038/nmeth.2307

19. Delaneau O. SHAPEIT. https://mathgen.stats.ox.ac.uk/genetics_software/shapeit/. Accessed October 5, 2021

20. Howie BN, Donnelly P, Marchini J (2009) A flexible and accurate genotype imputation method for the next generation of genome-wide association studies. PLoS Genet 5(6): e1000529. https://doi.org/10.1371/journal.pgen.1000529

21. Browning SR, Browning BL (2007) Rapid and accurate haplotype phasing and missing-data inference for whole-genome association studies by use of localized haplotype clustering. Am J Hum Genet 81(5):1084–1097. https://doi.org/10.1086/521987

22. Howie B, Fuchsberger C, Stephens M et al (2012) Fast and accurate genotype imputation in genome-wide association studies through pre-phasing. Nat Genet 44(8):955–959. https://doi.org/10.1038/ng.2354

23. Shi S, Yuan N, Yang M et al (2018) Comprehensive assessment of genotype imputation performance. Hum Hered 83(3):107–116. https://doi.org/10.1159/000489758

24. Roshyara NR, Horn K, Kirsten H et al (2016) Comparing performance of modern genotype imputation methods in different ethnicities. Sci Rep 6:34386. https://doi.org/10.1038/srep34386

25. Clarke L, Fairley S, Zheng-Bradley X et al (2017) The International Genome Sample Resource (IGSR): a worldwide collection of genome variation incorporating the 1000 genomes project data. Nucleic Acids Res 45(D1): D854–D8D9. https://doi.org/10.1093/nar/gkw829

26. The 1000 Genomes Project Consortium (2015) A global reference for human genetic variation. Nature 526(7571):68–74. https://doi.org/10.1038/nature15393

27. International Genome Sample Resource: Data Portal (2021) https://www.internationalgenome.org/data Accessed December 2, 2021

28. Huang GH, Tseng YC. Genotype imputation accuracy with different reference panels in admixed populations. BMC Proc. 2014;8 (Suppl 1 Genetic Analysis Workshop 18Vanessa Olmo):S64. https://doi.org/10.1186/1753-6561-8-S1-S64

29. Howie B, Marchini J. Impute2: analyzing whole chromosomes. https://mathgen.stats.ox.ac.uk/impute/impute_v2.html#whole_chroms. Accessed November 15, 2021

30. Luan JT, Teumer A, Zhao J, Fuchsberger C, Willer C (2012) IMPUTE2: 1000 genomes imputation cookbook. https://genome.sph.umich.edu/wiki/IMPUTE2:_1000_Genomes_Imputation_Cookbook. Accessed October 5, 2021

31. University of California Santa Cruz Genomics Institute: Cytoband. http://hgdownload.cse.ucsc.edu/goldenPath/hg19/database/cytoBand.txt.gz. Accessed November 20, 2021

32. Howie B, Marchini J. Impute2: details about 'info' metric. https://mathgen.stats.ox.ac.uk/impute/impute_v2.html#info_metric_details. Accessed October 5, 2021

33. Purcell S (2021) PLINK 1.9 input filtering. https://www.cog-genomics.org/plink/1.9/filter. Accessed November 15, 2021

34. Cordell HJ (2002) Epistasis: what it means, what it doesn't mean, and statistical methods to detect it in humans. Hum Mol Genet 11(20):2463–2468. https://doi.org/10.1093/hmg/11.20.2463

35. Panagiotou OA, Evangelou E, Ioannidis JP (2010) Genome-wide significant associations for variants with minor allele frequency of 5% or less – an overview: a HuGE review. Am J Epidemiol 172(8):869–889. https://doi.org/10.1093/aje/kwq234

36. Tryka KA, Hao L, Sturcke A et al (2014) NCBI's database of genotypes and phenotypes: dbGaP. Nucleic Acids Res 42(Database issue): D975–D979. https://doi.org/10.1093/nar/gkt1211

37. National Center for Biotechnology Information, National Library of Medicine: Database of Genotypes and Phenotypes (dbGaP). https://www.ncbi.nlm.nih.gov/gap/. Accessed December 5, 2021

38. Patterson N, Price AL, Reich D (2006) Population structure and eigenanalysis. PLoS Genet 2(12):e190. https://doi.org/10.1371/journal.pgen.0020190

39. Price AL, Patterson NJ, Plenge RM et al (2006) Principal components analysis corrects for stratification in genome-wide association studies. Nat Genet 38(8):904–909. https://doi.org/10.1038/ng1847

40. Auer PL, Lettre G (2015) Rare variant association studies: considerations, challenges and opportunities. Genome Med 7(1):16. https://doi.org/10.1186/s13073-015-0138-2

41. Cirillo E, Parnell LD, Evelo CT (2017) A review of pathway-based analysis tools that visualize genetic variants. Front Genet 8:174. https://doi.org/10.3389/fgene.2017.00174

42. Cooley PCC, Clark RF, Folsom RE (2014) Assessing gene-environment interactions in genome-wide association studies: statistical approaches, RTI Press research report series. Research Triangle Institute, Research Triangle Park

43. Lin WY, Huang CC, Liu YL et al (2018) Genome-wide gene-environment interaction analysis using set-based association tests. Front Genet 9:715. https://doi.org/10.3389/fgene.2018.00715

44. European Centre for Disease Prevention and Control (2021) West Nile virus infection. Annual epidemiological report for 2019. ECDC, Stockholm

45. Wojcik GL, Graff M, Nishimura KK et al (2019) Genetic analyses of diverse populations improves discovery for complex traits. Nature 570(7762):514–518. https://doi.org/10.1038/s41586-019-1310-4

46. Mills MC, Rahal C (2020) The GWAS diversity monitor tracks diversity by disease in real time. Nat Genet 52(3):242–243. https://doi.org/10.1038/s41588-020-0580-y

47. Peterson RE, Kuchenbaecker K, Walters RK et al (2019) Genome-wide association studies in ancestrally diverse populations: opportunities, methods, pitfalls, and recommendations. Cell 179(3):589–603. https://doi.org/10.1016/j.cell.2019.08.051

48. Atkinson EG, Maihofer AX, Kanai M et al (2021) Tractor uses local ancestry to enable the inclusion of admixed individuals in GWAS and to boost power. Nat Genet 53(2):195–204. https://doi.org/10.1038/s41588-020-00766-y

49. Cahill ME, Yao Y, Nock D et al (2017) West Nile virus seroprevalence, Connecticut, USA, 2000–2014. Emerg Infect Dis 23(4):708–710. https://doi.org/10.3201/eid2304.161669

50. Garcia MN, Hause AM, Walker CM et al (2014) Evaluation of prolonged fatigue post-West Nile virus infection and association of fatigue with elevated antiviral and proinflammatory cytokines. Viral Immunol 27(7):327–333. https://doi.org/10.1089/vim.2014.0035

51. Beloor J, Maes N, Ullah I et al (2018) Small interfering RNA-mediated control of virus replication in the CNS is therapeutic and enables natural immunity to West Nile virus. Cell Host Microbe 23(4):549–56 e3. https://doi.org/10.1016/j.chom.2018.03.001

52. Diamond MS (2009) Progress on the development of therapeutics against West Nile virus. Antivir Res 83(3):214–227. https://doi.org/10.1016/j.antiviral.2009.05.006

53. Ulbert S (2019) West Nile virus vaccines – current situation and future directions. Hum Vaccin Immunother 15(10):2337–2342. https://doi.org/10.1080/21645515.2019.1621149

54. Bai F, Thompson EA, Vig PJS et al (2019) Current understanding of West Nile virus clinical manifestations, immune responses, neuroinvasion, and immunotherapeutic implications. Pathogens 8(4). https://doi.org/10.3390/pathogens8040193

55. Bouaziz M, Ambroise C, Guedj M (2011) Accounting for population stratification in practice: a comparison of the main strategies dedicated to genome-wide association studies. PLoS One 6(12):e28845. https://doi.org/10.1371/journal.pone.0028845

56. Zhao H, Mitra N, Kanetsky PA et al (2018) A practical approach to adjusting for population stratification in genome-wide association studies: principal components and propensity scores (PCAPS). Stat Appl Genet Mol Biol 17(6). https://doi.org/10.1515/sagmb-2017-0054

57. Cross-Disorder Group of the Psychiatric Genomics Consortium (2019) Genomic relationships, novel loci, and pleiotropic mechanisms across eight psychiatric disorders. Cell 179(7):1469–82 e11. https://doi.org/10.1016/j.cell.2019.11.020

58. Wellcome Trust Case Control Consortium (2007) Genome-wide association study of 14,000 cases of seven common diseases and 3,000 shared controls. Nature 447(7145):661–678. https://doi.org/10.1038/nature05911

Chapter 18

Safety Procedures to Work with West Nile Virus in Biosafety Level 3 Facilities

Freedom M. Green and Shannon E. Ronca

Abstract

West Nile virus (WNV) can cause severe and sometimes fatal disease, but we do not have treatments or therapeutics to manage these outcomes. Since its introduction to the USA in 1999, WNV has been handled in a biosafety level 3 laboratory to decrease risk to researchers, requiring strict safety protocols and important considerations with planning experiments. Recent changes in US guidelines suggest that WNV can be handled at a lower biosafety level due to its endemicity in the USA and generally minor symptoms, but some research still requires the use of the agent at biosafety level 3. This chapter will briefly discuss the considerations of biosafety when working with WNV.

Key words Biosafety, Biosafety level, Risk group, West Nile virus, Arbovirus

1 Introduction to Biocontainment

West Nile virus (WNV) infections frequently lead to asymptomatic or subclinical infection, but in a subset of cases, this can progress to severe neurological disease, which can present as encephalitis, meningitis, and/or acute flaccid paralysis. Long-term complications of infection are common, with greater than 40% of infected patients reporting some form of sequelae, including memory loss and depression. There are no vaccines or antiviral drugs currently approved for use in human, leaving clinicians only with an option of supportive care. Together, these aspects of WNV infection have led to the recommendations in the USA that the virus be classified as a risk group level 3, requiring handling with biosafety level (BSL) 3 precautions [1].

When discussing agents, we refer to risk groups, while when discussing the facilities at which we handle these agents, we refer to the BSL. There are four levels of risk groups (Table 1) and, therefore, four levels of biosafety facilities (Table 2), with higher number

Fengwei Bai (ed.), *West Nile Virus: Methods and Protocols*,
Methods in Molecular Biology, vol. 2585, https://doi.org/10.1007/978-1-0716-2760-0_18,

Table 1
Classifications of risk group

Risk group	Per NIH guidelines	Per WHO guidelines
1	Not associated with disease in healthy adult humans	No or low individual or community risk, unlikely to cause human or animal disease
2	Associated with human disease that is rarely serious and for which preventive or therapeutic interventions are often available	Moderate individual risk; low community risk; causes human or animal disease but is unlikely to be a serious hazard to laboratory workers, the community, livestock, or the environment Laboratory exposures may cause serious infection, but effective treatment and preventive measures are available, and the risk of spread of infection is limited
3	Associated with serious or lethal human disease for which preventive or therapeutic interventions may be available (high individual risk but low community risk)	High individual risk, low community risk, usually causes serious human or animal disease but does not ordinarily spread from one infected individual to another Effective treatment and preventive measures are available
4	Agents likely to cause serious or lethal human disease for which preventive or therapeutic interventions are not usually available (high individual risk and high community risk)	High individual and community risk, causes serious human or animal disease and can be readily transmitted from one individual to another, directly or indirectly Effective treatment and preventive measures are not usually available

indicating the necessity of more precautions when handling agents. When assigning a BSL any biological agent, several aspects must be considered, including infectivity, severity of disease, transmissibility, treatability, the research being done, and the origin of the agent.

2   Biosafety Level 3 Requirements

All individuals that enter BSL-3 are required to complete training, but this training varies by institution. While the specifics may vary, the major requirements typically include a theoretical training module and hands-on training followed by a minimum number of mentored hours in the facility. These trainings are conducted as a collaborative effort between the Biosafety Officers, the Facility Directors, and the research teams. Open and honest communication is critical to successful training and operations of BSL-3 laboratories.

Table 2
Classifications of biosafety level and precautions

Biosafety level	Personal protective equipment	Facility requirements	Special considerations
1	Eye protection, gloves, and laboratory coats or gowns	Laboratory must contain doors, handwashing sink, laboratory bench, screened windows, and adequate lighting	None, standard microbiological processes are applied
2	Laboratory coats/gowns and gloves. Eye and respiratory protection to be worn as needed	Self-closing doors, biosafety cabinets, sealed windows, autoclave access, and sink near exit are required	Access is limited All procedures that may generate splashes or aerosols must be conducted in a biosafety cabinet
3	Protective solid front gowns, scrubs, or coveralls required. Two pairs of gloves and shoe covers to be worn when appropriate. Eye protection and respiratory protection, such as N95s or PAPRs	Laboratory is separate from access corridors and must contain two consecutive self-closing doors, biosafety cabinets, a hands-free sink near exit, and autoclave access. Seams, floors, walls, and ceilings must be sealed. Ducted mechanical air ventilation system with negative airflow into laboratory is required	Access is restricted to essential personnel All procedures with infectious material must be performed in a biosafety cabinet
4	Clothing to be changed prior to entry. Solid front gowns, scrubs, or coveralls and gloves are required in cabinet laboratory. Full-body, air-supplied, positive-pressure suit is required	Laboratory is accessed following entry sequence through airlock with airtight doors. Walls, floors, and ceilings are sealed to create a sealed internal shell. Dedicated nonrecirculating ventilation system and double-door pass-through autoclave are required	Daily inspection of containment and life support systems is required All procedures with viable infectious materials must be performed in a biosafety cabinet

All individuals associated with a BSL-3 are responsible for ensuring their own safety, as well as the safety of other facility users and the public by following existing guidelines. These guidelines address personal protective equipment (PPE), primary containment barriers, such as biosafety cabinet use, and secondary facility barriers and special precautions (Table 2). Selecting the appropriate personal protective equipment are vital to ensuring laboratory workers' safety. Selection of appropriate personal protective equipment should be decided after a detailed risk assessment of all agents a facility user may encounter within the facility. These assessments should begin by considering the quantity of WNV being used, what model system is in use, what techniques are being performed, and the experience level of the research team performing the experiments. Additional considerations will be applied as the details of the experiments are discussed with a team of biosafety and research professionals.

2.1 Considerations of Aerosolizing Techniques

Many common laboratory procedures produce aerosols, including pipetting. Smaller, respirable-sized particles can be suspended in the air for extended periods of time, causing infection through inhalation and contact with broken skin or exposed mucous membranes. Larger particles which settle more quickly may contaminate samples and work surfaces and enter the body through broken skin or contact of contaminated hands or other materials with mucous membranes. Laboratory-acquired infection through the inhalation of aerosolized viral particles is a significant risk that must be considered when evaluating the type of respiratory protection necessary for the proposed work, as well as using appropriate disinfectants on all surfaces to prevent contact contamination. Respiratory protection is especially vital for protection against accidents, such as open laboratory spills and laboratory equipment failures.

2.2 Animal Work

Working with animals and infectious agents, especially in high-containment settings, creates additional unique risks to the researcher. The most significant risks for laboratory-acquired infection of WNV when working with infected animals includes accidental inoculation through needle sticks or animal bites. Although updated guidelines suggest regular work with WNV can be done at BSL-2, the increased risks posed by animal work may dictate the continued use of certain BSL-3 animal-handling procedures or the decision to continue conducting animal work in the BSL-3. For example, the continued use of more stringent sharps precautions and avoiding direct contact with infected animals not under appropriate sedation are recommended to reduce the risk of infection via accidental sticks and animal bites. For example, this can include using forceps to move small rodents from their cages instead of picking them up with the hand. The specific precautions should always be discussed with experienced researchers, biosafety professionals, and the containment veterinarian.

2.3 Inactivation of BSL-3 Materials for Use at Lower Biosafety Levels

While the use of live agents requires a BSL-3 facility, some collected samples can be inactivated and removed from the facility for further processing. This inactivation applies to cell and animal samples, which can be inactivated in denaturing agents for DNA/RNA extractions, such as Trizol or DNA/RNA Shield, or in fixatives, such as formaldehyde, for removal of whole tissues for use in histology and immunofluorescence. The inactivation of any material should include testing to confirm the method is sufficient for inactivation for each agent. These protocols should be developed as a joint effort between researchers and biosafety professionals and should be fully validated before any materials are removed from the BSL-3.

3 Safe Handling of WNV

In the USA, the Biosafety in Microbiological and Biomedical Laboratories Manual (BMBL) acts as the handbook and guidance for determining the biosafety level necessary to work with infectious agents. Since the discovery of WNV through the fifth edition of the BMBL, WNV was classified as a risk group 3 agent, but the recent sixth edition of the manual has downgraded WNV to allow for handling in BSL-2 facilities [1]. This change was stated to be influenced by the endemicity of WNV in the USA and low instances of laboratory-acquired infections [1–4], but explicitly states that working with high titer virus or aerosolizing procedures creates justification for work in the BSL-3. As such, it is generally left to the Institutional Biosafety Committee (IBC) to help researchers apply the appropriate BSL. One could argue this low level of laboratory-acquired infections is, at least, in part due to the stringent safety guidelines of BSL-3 handling of agents.

As established research teams working with WNV already have all stocks in BSL-3 labs, there need to be clear, defined protocols or moving the stocks from BSL-3 to BSL-2, should this be approved by the biosafety professionals at the institution. These protocols should address how to assess and confirm that no other risk group 3 pathogens are present in the stocks.

Due to the former classification of WNV as a BSL-3 agent, facilities with a history of working with WNV will have stored WNV stocks, cultures, animal samples, and other related infectious materials in BSL-3 laboratories. As such, these facilities will need to work closely with their respective biosafety teams to adjust current guidelines for the future collection and storage of WNV materials at BSL-2 according to the updated recommendations found in the sixth edition of the BMBL released in June of 2020 [1].

4 Conclusions

Biosafety is at the heart of all arboviral research, protecting laboratory workers and the public from accidental release of agents and finding safe and creative ways to answer research questions. Researchers should develop strong relationships with their biosafety officers and IBC and approach unique research issues by providing detailing information with an open conversation. It is important to remember that what works at one institution may not be directly applicable to another institution. Adherence to the most comprehensive and up-to-date biosafety recommendations and institutional guidelines allows laboratory and clinical researchers to continue to expand the global understanding of the mechanisms and outcomes of WNV infections. Performing research safely ensures all researchers may continue to pursue strategies for treatment and management of WNV and other related infection.

References

1. Prevention CfDCa, Health NIo (2020) Biosafety in microbiological and biomedical laboratories, 6th edn. CDC, Atlanta

2. Wurtz N, Papa A, Hukic M, Di Caro A, Leparc-Goffart I, Leroy E, Landini MP, Sekeyova Z, Dumler JS, Badescu D, Busquets N, Calistri A, Parolin C, Palu G, Christova I, Maurin M, La Scola B, Raoult D (2016) Survey of laboratory-acquired infections around the world in biosafety level 3 and 4 laboratories. Eur J Clin Microbiol Infect Dis 35(8):1247–1258. https://doi.org/10.1007/s10096-016-2657-1

3. From the Centers for Disease Control and Prevention (2003) Laboratory-acquired West Nile virus infections—United States, 2002. JAMA 289(4):414–415

4. Centers for Disease Control and Prevention (2002) Laboratory-acquired West Nile virus infections—United States, 2002. MMWR Morb Mortal Wkly Rep 51(50):1133–1135

Chapter 19

Development of Antibody-Based Therapeutics Against West Nile Virus in Plants

Haiyan Sun, Josh Lesio, and Qiang Chen

Abstract

Since its discovery in 1937 in the West Nile district of Uganda, West Nile virus (WNV) has been one of the leading causes of mosquito-transmitted infectious diseases (Smithburn, Burke, Am J Trop Med 20:22, 1940). Subsequently, it spread to Europe, Asia, Australia, and finally North America in 1999 (Sejvar, Ochsner 5(3):6–10, 2003). Worldwide outbreaks have continued to increase since the 1990s (Chancey et al, Biomed Res Int 2015:376230, 2015). According to the Center for Disease Control and Prevention, more than 51,000 cases of WNV infection and nearly 2400 cases of WNV-related death were reported in the USA from 1999 to 2019. The estimated economic impact of WNV infections is close to 800 million dollars in the USA from 1999 to 2012 (Barrett, Am J Trop Med Hyg 90:389, 2014).

Key words West Nile virus, Monoclonal antibody (mAb), Plant-derived antibody, Plant expression system, Plant-made biologics

1 West Nile Virus Overview

While roughly 80% of WNV-infected individuals are asymptomatic, the remaining 20% have mild symptoms such as fever, headache, fatigue, or muscle pain. Approximately 1 in 150 infections lead to life-threatening West Nile neuroinvasive disease (WNND). WNND includes encephalitis (WNE), meningitis (WNM), and acute flaccid paralysis (AFP) depending on the site of infection [1, 2]. The risk of WNND is increased by old age and a history of chronic diseases such as cancer, diabetes, or hypertension [2]. Among animals, horses are the most WNV-impacted mammals with higher infection rates and have more severe symptoms than humans. Without veterinary vaccines, the fatality rate is almost one-third with 40% of surviving horses suffering from neurological sequelae [3].

WNV circulation is maintained by mosquitos and wild birds. In the USA, WNV has been detected in several dozen mosquito species and more than 300 bird species [2]. Since its discovery in

Fengwei Bai (ed.), *West Nile Virus: Methods and Protocols*,
Methods in Molecular Biology, vol. 2585, https://doi.org/10.1007/978-1-0716-2760-0_19,

1937, diverse strains of WNV have been isolated in different geographical regions. Currently, WNV is composed of up to nine lineages, with genome sequence variation of at least 20%–25% between lineages [4, 5]. Lineages 1 and 2 cause the majority of WNV outbreaks, with 1 being the most widespread [5]. Lineage 1 is further classified into lineage 1a, 1b, and 1c by geographic location of discovery—lineage 1a, the globally spread strain, has been reported in Africa, Europe, and North America; lineage 1b, referred to as Kunjin strain, was only reported in Australia; and lineage 1c, was identified in India and sometimes classified as Lineage 5 [5]. Lineage 2 was originally reported in Africa and later found in Southern and Eastern Europe. Both lineages 1 and 2 contain strains that can cause neuroinvasive infections in humans [5]. Compared to lineages 1 and 2, the other lineages are less widely spread. Lineage 3 was found only in the Czech Republic and lineage 4 was reported in Russia [4]. Lineage 7, or the Koutango virus, was previously considered a different virus and was found in ticks and rodents [5]. Lineages 6, 8, and 9 are putative lineages reported in Spain, Senegal, and Austria, respectively [5].

WNV belongs to the family of *Flaviviridae* from the *Flavivirus* genus, which also includes other well-known flaviviruses such as dengue virus (DENV), Zika virus (ZIKV), and yellow fever virus (YFV). Similar to other flaviviruses, WNV is an enveloped RNA virus with icosahedral symmetry [6]. The virus structure contains a host membrane-derived envelope and a nucleocapsid core [6]. Two structural proteins, the envelope protein (E) and precursor membrane protein (prM) or its processed product, the membrane protein (M), cover the surface of the viral envelope and form a relatively smooth protein shell. The nucleocapsid core consists of the capsid protein (C) and the single-stranded 11 kb RNA genome. The WNV genome encodes the three structural proteins (E, prM/M, and C) and seven nonstructural proteins (NS1, NS2A, NS2B, NS3, NS4A, NS4B, and NS5). The functions of the nonstructural proteins are not fully studied, but they are known to play important roles in virus RNA replication, protein synthesis, and virus replication complex assembly in the infected host cells [7, 8]. The capsid protein is the first viral protein to be synthesized in the infected cells, indicating the capsid protein is critical for virus replication and packaging, and may influence virulence and pathogenesis through interactions with host proteins [7, 9]. The prM and E protein form heterodimers in the immature WNV particles [6, 10]. Low pH exposure during the secretory process triggers a conformation change of the E protein and leads to the cleavage of prM by protease furin to yield the mature M protein and infectious WNV virions [10]. E homodimers in the mature WNV are ready to mediate membrane fusion. However, it was shown that the processed pr peptide remains associated with the virus at low pH and prevents premature E protein-mediated membrane fusion before exocytosis of the virus [10]. The glycosylated E protein is involved in receptor binding on the host cell surface and facilitates the virus entry through membrane fusion. The crystal structure of WNV E protein revealed that it has three distinct domains: a beta barrel-shaped domain I (DI), an elongated domain II with a highly conserved fusion loop (DII), and the C-terminal immunoglobulin-like domain III (DIII) [11]. The WNV E protein has been the main target for neutralizing antibodies, many of which specifically recognize epitopes in DIII [12]. In addition to the E protein, neutralizing antibodies have been identified targeting prM [13] or even the nonstructural proteins [14, 15].

2   Neutralizing Antibodies as WNV Therapeutics

The majority of WNV-infected people are either asymptomatic or have mild symptoms often resolved without medical attention. However, ~1% of infected individuals have more severe symptoms leading to WNND, 10% of which die each year [16]. Currently, there are no licensed therapeutics or vaccines available for WNV infection in humans [17]. All WNV treatments are supportive care such as intravenous fluids and pain medications.

Since the first approval of monoclonal antibody (mAb), muromonab-CD3 in 1985, which treats acute rejection in organ transplants [18], more than 100 antibody drugs have been approved by the US Food and Drug Administration (FDA) as of March 2021 [19]. These antibody drugs provide new treatments for diverse human diseases including cancers, infectious diseases, chronic inflammatory diseases, and neurological diseases.

Recently, the FDA has issued emergency use authorization (EUA) of several mAbs for SARS-CoV-2 treatment, highlighting the significance of antibody drugs in fighting infectious diseases. Currently, there have been six mAbs approved by the FDA for the prevention or treatment of infectious diseases [19]. In 1998, Palivizumab, the first mAb against infectious disease, was licensed for the prevention of respiratory syncytial virus (RSV) infection, the most common cause for severe bronchiolitis in young children [20]. Palivizumab prevents RSV entry into host cells by binding to the RSV envelop fusion protein and inhibiting membrane fusion [21]. Two mAbs, Raxibacumab and Obiltoxaximab, were licensed for the treatment of inhalational anthrax, both of which target the bacterial antigen responsible for cell binding during infection [22, 23]. In 2018, Ibalizumab, a humanized immunoglobulin G4, was approved for clinical management of human immunodeficiency virus (HIV)-1 infection with multidrug resistance [24]. As a CD4-directed post-attachment inhibitor, this mAb binds to the

CD4 T cells and blocks the conformation changes required for HIV-1 entry [24]. In 2020, two antibody drugs (Inmazeb and Ebanga) were licensed for the treatment of Ebola virus (EBOV) infections. Inmazeb is a combination of three mAbs: atoltivimab, maftivimab, and odesivimab, all three of which can bind to the EBOV glycoprotein simultaneously to prevent virus entry [25]. Ebanga contains one mAb called ansuvimab, which also binds the EBOV glycoprotein and blocks its interaction with the host cell receptor [26]. In addition to blocking virus attachment or membrane fusion to prevent virus entry, increasing evidence shows that antibodies can provide significant therapeutic effects through fragment crystallizable region (Fc)-mediated effector functions, such as complement-dependent cytotoxicity (CDC), antibody-dependent cell cytotoxicity (ADCC), and antibody-dependent cellular phagocytosis (ADCP) [27–29]. For example, two of the mAbs in the EBOV drug, Inmazeb and Ansuvimab, can induce ADCC in addition to blocking virus entry [30, 31], indicating the importance of effector functions in the treatment of EBOV infection.

Like other flaviviruses, most identified neutralizing antibodies against WNV target the E protein. Various epitopes have been mapped to all three domains of E protein from neutralizing antibodies isolated from human WNV patients or mice challenged with WNV infection. The most promising WNV antibody, E16, was originally identified by screening a large panel of mouse mAbs generated against soluble WNV E protein [12]. In vitro plaque reduction assays demonstrated that E16 was able to potently neutralize genetically diverse strains of WNV from lineages 1 and 2 [12]. Both the mouse E16 and humanized E16 could protect mice from WNV lethal challenge with one single dose treatment postinfection. Mutagenesis and yeast surface display of DIII of WNV E protein indicated that E16 binds to the lateral ridge in DIII [12]. The crystal structure of the E16 Fab region in complex with WNV DIII further revealed that E16 interacts with a conformational epitope containing 16 amino acid residues in DIII. These 16 amino acids come from four discontinuous peptide segments of DIII, including the N terminal region and three strand-connecting loops, which together form a uniquely conserved surface patch in the lateral ridge of WNV DIII [32]. Studies using complement-deficient or Fc gamma receptor (FcγR)-knockout mice indicate that Fc-mediated effector functions contribute to the overall therapeutic effects of E16 [12]. The promising results from in vitro and in vivo studies led to the Phase I/II clinical trials using humanized E16 (Hu-E16) for the treatment of severe WNV infection [33]. However, the trials were terminated early due to difficulty in enrolling enough qualified WNV patients.

Potent neutralizing antibodies have also been isolated from WNV-infected human patients. CR4374 and CR4353, two of the antibodies targeting DIII, showed a protective effect in mice when challenged with a lethal dose of WNV [34]. Interestingly, two WNV specific human antibodies, CR4348 and CR4354, do not bind to the recombinant E protein, but strongly neutralize WNV in vitro and protect mice from lethal challenges in vivo [35]. Mutagenesis studies and structural analysis suggested that CR4348 and CR4354 may interact with the dimer interface in DII, and the hinge between DI and DII, respectively [35]. More recently, WNV-86, a new human antibody isolated from serum samples of the 2012 WNV outbreak, demonstrated higher potency than E16 and protected mice from lethal challenge with one single dose [36]. Distinct from E16, WNV-86 recognizes an epitope in DII and preferentially neutralizes the mature virions lacking the uncleaved prM [36].

Antibody-dependent enhancement (ADE) of infection has always been a serious concern in the development of antibody therapeutics against flavivirus infections [27]. ADE occurs when the virus binds to non−/sub-neutralizing antibodies or neutralizing antibodies at sub-neutralizing concentrations, and subsequently, the antibody-virus complex enters the FcγR-expressing myeloid cells through FcγR-mediated endocytosis. This phenomenon was first observed clinically in patients with secondary infection of a heterologous DENV serotype that led to dengue hemorrhagic fever (DHF) and dengue shock syndrome (DSS) [37]. Although not fully understood, in vitro studies have confirmed ADE's existence in a variety of viral infections including DENV, WNV, ZIKV, and EBOV [38, 39]. More recently, the well-conserved flavivirus NS1 protein has become a prominent target for vaccine and drug development as the nonstructural proteins present a very low risk for ADE. Early this year, two murine antibodies against NS1 protein, reported by two different research groups, 1G5.3 and 2B7, showed high binding affinity to various flavivirus NS1 proteins and strongly inhibit NS1-induced endothelial dysfunction [14, 15]. Structural analysis indicated that 2B7 inhibits two domains of DENV NS1 protein simultaneously to prevent initial binding to endothelial cells and downstream events [15]. Crystal structures of 1G5.3 complexed with DENV or ZIKV NS1 C-terminal fragment suggested that 1G5.3 protects from flavivirus infections through blocking early interactions of NS1 with cell surface receptors [14]. In vivo experiments demonstrated that 1G5.3 improves survival rates in mouse models challenged with lethal doses of DENV, ZIKV, or WNV, respectively. The therapeutic effects of both 1G5.3 and 2B7 are independent of Fc effector functions [14, 15]. These studies provide new insights into the development of broadly protective antibodies against multiple flavivirus pathogens.

3   Plant System for Producing Recombinant Proteins

Plants have been proved to be a useful system for the production of recombinant therapeutic proteins in the last two decades. Plant-based expression systems possess unique advantages for the development and production of recombinant therapeutic proteins compared to traditional mammalian expression systems [27, 40]. Plant-based systems are attractive not only because of their low growth cost and scalability potential, but also plants do not carry human pathogens and therefore pose no risk for contamination [41]. In addition, recent advances in "deconstructed" viral vector systems allow for rapid production of recombinant proteins with high yield in less than 2 weeks [42, 43]. Furthermore, as plants have a small repertoire of glycoenzymes and are amenable to glycoengineering, plant-based expression systems may provide the necessary glycosylation patterns to reduce ADE in developing antibody therapeutics [27, 44].

In the 1970s, it was discovered that *Agrobacterium tumefaciens* contains a tumor-inducing plasmid that could transfer and integrate part of its DNA into the plant genome [45]. This discovery led to the rapid development of transgenic plants with stable expression of genes of interest (GOI) in the early years. However, developing transgenic plants is often time-consuming and has poor protein expression for GOI [46]. More recently, there have been great successes in the development of plant viral expression vectors facilitating the production of recombinant proteins with transient expression in plants. Several plant viruses have been used to construct plant expression vectors, such as potato virus X (PVX), tobacco mosaic virus (TMV), cowpea mosaic virus (CPMV), cucumber mosaic virus (CMV), geminivirus, and cauliflower mosaic virus (CaMV) [46, 47]. Along with the advances in agroinfiltration techniques, the combination of plant viral vectors and *Agrobacterium* has made it possible to deliver GOI and produce recombinant proteins with high yield in a short time period [48, 49].

The magnICON system was the first rationally designed deconstructed viral vector for high-yield recombinant protein production in plants [50]. It splits a hybrid tobamovirus genome into three components which are transformed into *A. tumefaciens* separately and co-infiltrated into the same plants from a mixture of the three *Agrobacterium* cultures. In the magnICON system, the 5′ module contains the TMV-based elements necessary for replication, while the 3′ module has the GOI, and a recombinase module carries a streptomyces phage 31 integrase [50]. Once all three modules are delivered into the plants, the integrase ligates the 5′ module and the 3′module together to form a complete replicon, allowing the replication of GOI RNA and production of the protein of interest. The magnICON system has been successfully used for small- and large-scale production of various pharmaceutical proteins [50].

Another popularly used plant viral vector is the geminivirus-based expression system. Different from the magnICON system which was designed based on single-stranded RNA viruses, the geminivirus-based expression system takes advantage of the single-stranded circular DNA genome of geminiviruses [51]. The geminivirus genome contains a long intergenic region (LIR), short intergenic region (SIR), and sequences coding for the replication-associated proteins (Rep/RepA). A bean yellow dwarf virus (BeYDV)-based expression vector is a good example of geminiviral-based deconstructed expression vectors [52]. It includes a Rep/RepA cassette and a SIR flanked by two LIRs (LSL) with a cloning site for the GOI in the LSL region [53]. The Rep/RepA expression cassette can also be provided in trans [51]. Once they are delivered to plant cells, the Rep/RepA proteins circularize the DNA between the two LIRs and initiate the replication of LSL. The geminivirus-based system allows the expression of a single large GOI or multiple small GOIs in one single viral vector, while the expression level is regulated by Rep/RepA expression [46, 49].

4 Plant-Produced mAb-Based Therapeutics Against WNV

Over the past decades, WNV has spread throughout the world with no licensed WNV antiviral treatment for human use. The reality of WNV epidemics requires rapid development and scalable production of antibody therapeutics in a cost-efficient manner. Plant expression systems have shown promising attributes for producing pharmaceutical proteins. Our lab has been long interested in the development of antibody therapeutics and vaccines for infectious diseases in plants to improve their efficacy and safety [54–61]. Here we summarize the development and functional analysis of plant-produced E16 (pHu-E16) and its variants. The Hu-E16 exhibits high neutralization potency against WNV infection and was able to protect mice from WNV lethal challenge [12]. Using *Nicotiana benthamiana* plants, we demonstrated that pHu-E16 can be robustly expressed in 8 days post infiltration (dpi) [62]. pHu-E16 can be easily purified with >95% homogeneity, and the production method can be scaled up without compromising the rate of pHu-E16 accumulation and recovery [62–64]. pHu-E16 recognizes and binds to the WNV E protein DIII identically as Hu-E16 is produced from mammalian cell culture (mHu-E16). In vitro neutralization assay suggested that pHu-E16 neutralizes WNV reporter virus particles equivalently to that of mHu-E16. However, pHu-E16 binds to C1q with lower affinity than mHu-E16, and its neutralization potency was not increased in the presence of C1q, probably due to the different N-glycosylation in pHu-E16 [62]. Nevertheless, pHu-E16 protects against lethal WNV infection indistinguishably from that of mHu-E16 in mouse models. These studies provide proof of concept for the production of antibody therapeutics in plant expression systems. To explore alternative cost-effective and scalable production of therapeutic mAbs in plants, we chose lettuce (*Lactuca sativa*), a fast-growing, inexpensive plant with a lower level of toxic secondary metabolites, to produce Hu-E16 (LHu-E16) using the BeYDV-based geminivirus vector [65, 66]. The accumulation of LHu-E16 in lettuce leaves was about 0.27 mg/g leaf fresh weight (LFW) within 4 dpi, which is 4 days shorter than that of pHu-E16 produced in *N. benthamiana* [65]. LHu-E16 has the same binding affinity to WNV E protein or DIII and neutralizes WNV infection very similar to mHu-E16.

Recombinant proteins produced in wild-type plants contain plant-specific glycans that are different from proteins produced in mammalian cell culture. To eliminate the possible undesirable immune responses induced by plant-specific glycans, Hu-E16 and a variant with single-chain variable fragment fused to the heavy-chain constant domains of human IgG (E16scFv-CH) were produced in glycoengineered *N. benthamiana* plants (ΔXF) of which the endogenous β1,2-xylosyltransferase and α1,3-fucosyltransferase were silenced by RNAi [67, 68]. Both ΔXFpE16 and ΔXFpE16scFv-CH were expressed with high protein levels and assembled correctly similar to that of pHu-E16 in wild-type *N. benthamiana* [69]. Glycan analysis showed that both ΔXFpE16 and ΔXFpE16scFv-CH displayed a mammalian-type glycosylation profile without any trace of plant-specific glycans. Antigen-binding assays demonstrated that both ΔXFpE16 and ΔXFpE16scFv-CH bind to WNV E protein DIII in a manner almost identical to that of mHu-E16 [69]. Neutralization assay indicated that ΔXFpE16 and ΔXFpE16scFv-CH neutralize WNV RVP slightly better than or equivalent to mHu-E16, respectively. In the presence of C1q, ΔXFpE16scFv-CH but not ΔXFpE16 had a modest left shift of the neutralization curve similar to mHu-E16, which may be due to their slight differences in N-glycosylation [69]. Both ΔXFpE16 and ΔXFpE16scFv-CH exhibited preventive and therapeutic efficacy indistinguishable from mHu-E16. These studies demonstrate the feasibility of developing anti-WNV mAb therapeutics and their single-chain variants in glycoengineered plants with low cost and better safety with plant-specific glycans eliminated.

WNV infection can be neuroinvasive and cause life-threatening WNND in both human and animal cases. Typical mAb therapeutics would have very limited efficacy in treating WNND due to their

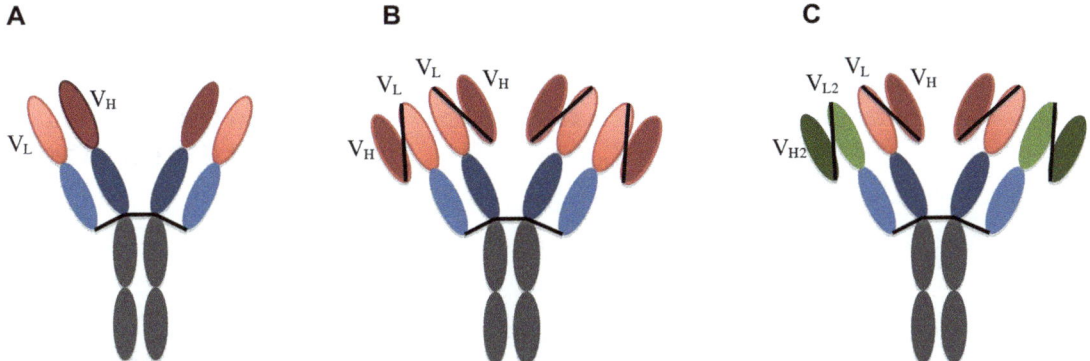

Fig. 1 Tetravalent E16 and bifunctional E16 design. (**a**) E16 mAb. (**b**) Tetravalent E16. (**c**) Bifunctional E16 that are designed to enhance its ability to cross the blood-brain barrier. V_L: variable region of light chain; V_H: variable region of heavy chain; V_L/V_H: the first pair of antigen-binding sites that bind and neutralize WNV. V_{L2}/V_{H2}: a second pair of antigen-binding sites that bind to surface receptors on the endothelial cells of the BBB. This may enhance the permeability of the bifunctional antibody across the BBB via receptor-mediated transcytosis

inability to cross the blood-brain barrier (BBB). More complexed antibody variants, such as bispecific antibodies, have been shown to have higher efficacy as one functionality could be designed to facilitate transport across the BBB [70, 71]. To explore the possibility of using plant systems to produce more complexed antibody variants, we designed a tetravalent Hu-E16 variant (tetra pHu-E16) assembled from pE16scFv-CH and a second pE16scFv fused to the human IgG light-chain (LC) constant region (Fig. 1) [72]. In addition, we expressed other Hu-E16 variants in plants with various combinations of human IgG HC and LC components to evaluate how the differences in N-glycosylation impact pHu-E16 variant assembly and function. Our results revealed that all the Hu-E16 variants we tested were expressed and correctly assembled in plants similarly to pHu-E16. Detailed glycan analysis of all variants indicated that proper pairing of HC and LC was crucial for the complete N-glycosylation of antibodies in both plant and animal cells. N-glycosylation affects antibody binding to C1q and Fcγ receptors. Not surprisingly, surface plasmon resonance (SPR) analysis revealed that tetra pHu-E16 and other variants exhibit differential binding affinity to C1q and various Fcγ receptors. More importantly, unlike mHu-E16, none of the plant-produced Hu-E16 variants displayed any ADE activity in CD32A-expressing cells infected with WNV (Fig. 2). Both in vitro and in vivo studies suggested that tetra pHu-E16 neutralizes WNV infection and protects mice from WNV lethal challenge in equivalent to mHu-E16 (Fig. 3). These studies indicate plant expression systems are capable of producing more complexed antibody therapeutics against WNV infection without the risk of ADE. While the role of ADE during

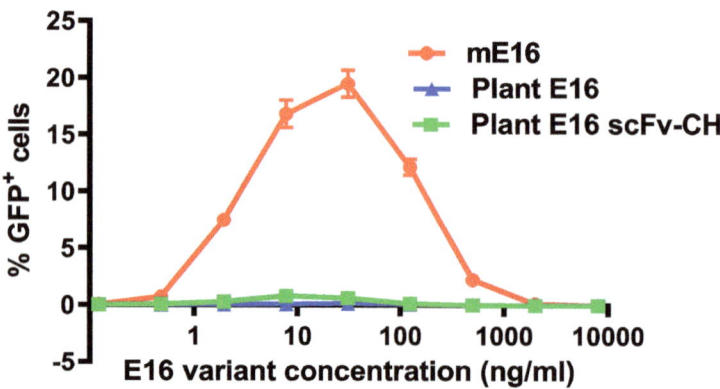

Fig. 2 Antibody-dependent enhancement of E16 variants. Plant-produced E16 (Plant E16), plant-made single-chain E16 (E16 scFv-CH), or mammalian cell-made E16 (mE16) was serially diluted and mixed with GFP pseudo-infectious WNV replicon particles (RVP). The mAb and RVP mixture was then added to FcγR-expressing K562 cells. Forty-eight hours later, cells were analyzed by flow cytometry GFP expression. (From Ref. [72] with permission)

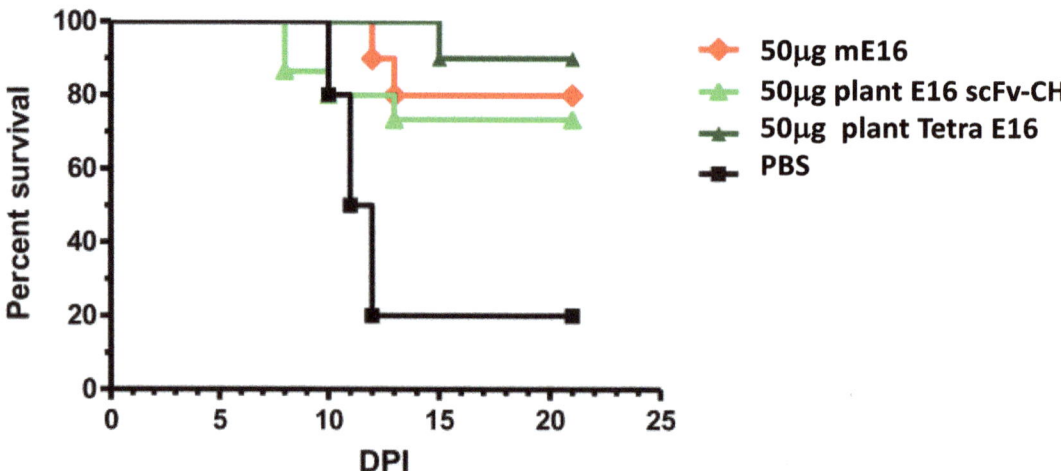

Fig. 3 Plant-made E16 mAb variants protected mice against a lethal challenge of WNV infection. Wild-type C57BL/6 mice were infected with 10^2 PFU of WNV. A single 50-μg dose of plant-made single-chain E16 (E16 scFv-C$_H$) or tetra E16 was given to mice via an intraperitoneal route 4 days after infection. Mammalian cell-made E16 (mE16) was used as a positive control. Survival data was pooled from several independent experiments ($n > 10$ mice per dose) and analyzed by the log-rank test. (From Ref. [72] with permission)

WNV pathogenesis is uncertain, our results suggest that plant glycoengineering is a promising approach in eliminating the risk of ADE associated with mAb-based therapeutics against ADE-prone viruses such as dengue virus [73].

5 Conclusions

Plant-based expression platforms offer multiple advantages beyond the traditional benefits of low production cost, facile scalability, and low risk of contamination by animal pathogens. The development of deconstructed plant viral expression vectors has allowed the development and production of anti-WNV mAbs at unprecedented speed to combat potential sudden outbreaks in different parts of the world. Glycoengineering allows plants to make WNV mAbs with tailor-made human N-glycans on demand to eliminate ADE and potentially increase beneficial effector functions. The testing of plant-made mAbs against Ebola (ZMapp) and HIV (2G12) in human clinical trials along with the approval of the first plant-made biologic (taliglucerase alfa) by the FDA demonstrates the safety of plant-made mAbs in humans and has cleared the regulatory pathway for approving future plant-derived mAb drugs against WNV [74–76].

Acknowledgments

The work from the Chen laboratory reported in this chapter was supported in part by grants from the National Institute of Allergy and Infectious Diseases (NIAID) # U01 AI075549 and # R33AI101329 to QC.

References

1. Alli A, Ortiz JF, Atoot A, Atoot A, Millhouse PW (2021) Management of West Nile encephalitis: an uncommon complication of West Nile virus. Cureus 13(2):e13183. https://doi.org/10.7759/cureus.13183

2. Petersen LR, Brault AC, Nasci RS (2013) West Nile virus: review of the literature. JAMA 310(3):308–315

3. Aharonson-Raz K, Lichter-Peled A, Tal S, Gelman B, Cohen D, Klement E, Steinman A (2014) Spatial and temporal distribution of West Nile virus in horses in Israel (1997–2013) – from endemic to epidemics. PLoS One 9(11):e113149. https://doi.org/10.1371/journal.pone.0113149

4. Chancey C, Grinev A, Volkova E, Rios M (2015) The global ecology and epidemiology of West Nile virus. Biomed Res Int 2015:376230. https://doi.org/10.1155/2015/376230

5. Fall G, Di Paola N, Faye M, Dia M, Freire CCM, Loucoubar C, Zanotto PMA, Faye O, Sall AA (2017) Biological and phylogenetic characteristics of West African lineages of West Nile virus. PLoS Negl Trop Dis 11(11):e0006078. https://doi.org/10.1371/journal.pntd.0006078

6. Mukhopadhyay S, Kim BS, Chipman PR, Rossmann MG, Kuhn RJ (2003) Structure of West Nile virus. Science 302(5643):248

7. Brinton MA (2013) Replication cycle and molecular biology of the West Nile virus. Viruses 6(1):13–53. https://doi.org/10.3390/v6010013

8. Tseng AC, Nerurkar VR, Neupane KR, Kae H, Kaufusi PH (2021) Potential dual role of West Nile virus NS2B in orchestrating NS3 enzymatic activity in viral replication. Viruses 13(2). https://doi.org/10.3390/v13020216

9. Kobayashi S, Yoshii K, Phongphaew W, Muto M, Hirano M, Orba Y, Sawa H, Kariwa H (2020) West Nile virus capsid protein inhibits autophagy by AMP-activated protein kinase degradation in neurological disease development. PLoS Pathog 16(1):e1008238.

https://doi.org/10.1371/journal.ppat.1008238

10. Pierson TC, Diamond MS (2012) Degrees of maturity: the complex structure and biology of flaviviruses. Curr Opin Virol 2(2):168–175. https://doi.org/10.1016/j.coviro.2012.02.011

11. Nybakken GE, Nelson CA, Chen BR, Diamond MS, Fremont DH (2006) Crystal structure of the West Nile virus envelope glycoprotein. J Virol 80(23):11467–11474. https://doi.org/10.1128/JVI.01125-06

12. Oliphant T, Engle M, Nybakken GE, Doane C, Johnson S, Huang L, Gorlatov S, Mehlhop E, Marri A, Chung KM, Ebel GD, Kramer LD, Fremont DH, Diamond MS (2005) Development of a humanized monoclonal antibody with therapeutic potential against West Nile virus. Nat Med 11(5):522–530. https://doi.org/10.1038/nm1240

13. Calvert AE, Kalantarov GF, Chang GJ, Trakht I, Blair CD, Roehrig JT (2011) Human monoclonal antibodies to West Nile virus identify epitopes on the prM protein. Virology 410(1):30–37. https://doi.org/10.1016/j.virol.2010.10.033

14. Naphak Modhiran Song H, Liu L, Bletchly C, Brillault L, Amarilla AA, Xu X, Qi J, Chai Y, Cheung STM, Traves R, Setoh YX, Bibby S, Scott CAP, Freney ME, Newton ND, Khromykh AA, Chappell KJ, Muller DA, Stacey KJ, Landsberg MJ, Shi Y, Gao GF, Young PR, Watterson D (2021) A broadly protective antibody that targets the flavivirus NS1 protein. Science 371(6525):190–194

15. Biering SB, Akey DL, Wong MP, Brown WC, Lo NTN, Puerta-Guardo H, Tramontini Gomes de Sousa F, Wang C, Konwerski JR, Espinosa DA, Bockhaus NJ, Glasner DR, Li J, Blanc SF, Juan EY, Elledge SJ, Mina MJ, Beatty PR, Smith JL, Harris E (2021) Structural basis for antibody inhibition of flavivirus NS1-triggered endothelial dysfunction. Science 371(6525):194–200. https://doi.org/10.1126/science.abc0476

16. Bai F, Thompson EA, Vig PJS, Leis AA (2019) Current understanding of West Nile virus clinical manifestations, immune responses, neuroinvasion, and immunotherapeutic implications. Pathogens 8(4). https://doi.org/10.3390/pathogens8040193

17. Saiz JC (2020) Animal and human vaccines against West Nile virus. Pathogens 9(12). https://doi.org/10.3390/pathogens9121073

18. Smith SL (1996) Ten years of orthoclone OKT3 (muromonab-CD3): a review. J Transpl Coord 6(3):109–119; quiz 120-101. https://

doi.org/10.7182/prtr.1.6.3.8145l3u185493182

19. Antibody Society (2022) Antibody therapeutics approved or in regulatory review in the EU or US. https://www.antibodysociety.org/resources/approved-antibodies/

20. Diseases AAoPCoI (2014) Updated guidance for palivizumab prophylaxis among infants and young children at increased risk of hospitalization for respiratory syncytial virus infection. Pediatrics 134:e620–e638

21. Huang K, Incognito L, Cheng X, Ulbrandt ND, Wu H (2010) Respiratory syncytial virus-neutralizing monoclonal antibodies motavizumab and palivizumab inhibit fusion. J Virol 84(16):8132–8140. https://doi.org/10.1128/JVI.02699-09

22. Migone TS, Subramanian GM, Zhong J, Healey LM, Corey A, Devalaraja M, Lo L, Ullrich S, Zimmerman J, Chen A, Lewis M, Meister G, Gillum K, Sanford D, Mott J, Bolmer SD (2009) Raxibacumab for the treatment of inhalational anthrax. N Engl J Med 361(2):135–144. https://doi.org/10.1056/NEJMoa0810603

23. Yamamoto BJ, Shadiack AM, Carpenter S, Sanford D, Henning LN, O'Connor E, Gonzales N, Mondick J, French J, Stark GV, Fisher AC, Casey LS, Serbina NV (2016) Efficacy projection of obiltoxaximab for treatment of inhalational anthrax across a range of disease severity. Antimicrob Agents Chemother 60(10):5787–5795. https://doi.org/10.1128/AAC.00972-16

24. Beccari MV, Mogle BT, Sidman EF, Mastro KA, Asiago-Reddy E, Kufel WD (2019) Ibalizumab, a novel monoclonal antibody for the Management of multidrug-resistant HIV-1 infection. Antimicrob Agents Chemother 63(6). https://doi.org/10.1128/AAC.00110-19

25. Atoltivimab, maftivimab, and odesivimab-ebgn (2021) Am J Health Syst Pharm 78(4):279–281. https://doi.org/10.1093/ajhp/zxaa404

26. Lee A (2021) Ansuvimab: first approval. Drugs. https://doi.org/10.1007/s40265-021-01483-4

27. Sun H, Chen Q, Lai H (2017) Development of antibody therapeutics against flaviviruses. Int J Mol Sci 19(1). https://doi.org/10.3390/ijms19010054

28. Su B, Dispinseri S, Iannone V, Zhang T, Wu H, Carapito R, Bahram S, Scarlatti G, Moog C (2019) Update on Fc-mediated antibody functions against HIV-1 beyond neutralization.

Front Immunol 10:2968. https://doi.org/10.3389/fimmu.2019.02968

29. van Erp EA, Luytjes W, Ferwerda G, van Kasteren PB (2019) Fc-mediated antibody effector functions during respiratory syncytial virus infection and disease. Front Immunol 10:548. https://doi.org/10.3389/fimmu.2019.00548

30. Pascal KE, Dudgeon D, Trefry JC et al (2018) Development of clinical-stage human monoclonal antibodies that treat advanced Ebola virus disease in nonhuman primates. J Infect Dis 218:S612–S626. https://doi.org/10.1093/infdis/jiy285

31. Corti D, Misasi J, Mulangu S, Stanley DA, Kanekiyo M, Wollen S, Ploquin A, Doria-Rose NA, Staupe RP, Bailey M, Shi W, Choe M, Marcus H, Thompson EA, Cagigi A, Silacci C, Fernandez-Rodriguez B, Perez L, Sallusto F, Vanzetta F, Agatic G, Cameroni E, Kisalu N, Gordon I, Ledgerwood JE, Mascola JR, Graham BS, Muyembe-Tamfun JJ, Trefry JC, Lanzavecchia A, Sullivan NJ (2016) Protective monotherapy against lethal Ebola virus infection by a potently neutralizing antibody. Science 351(6279):1339–1342. https://doi.org/10.1126/science.aad5224

32. Nybakken GE, Oliphant T, Johnson S, Burke S, Diamond MS, Fremont DH (2005) Structural basis of West Nile virus neutralization by a therapeutic antibody. Nature 437(7059):764–769. https://doi.org/10.1038/nature03956

33. Diamond MS (2009) Progress on the development of therapeutics against West Nile virus. Antivir Res 83(3):214–227. https://doi.org/10.1016/j.antiviral.2009.05.006

34. Throsby M, Geuijen C, Goudsmit J, Bakker AQ, Korimbocus J, Kramer RA, Clijsters-van der Horst M, de Jong M, Jongeneelen M, Thijsse S, Smit R, Visser TJ, Bijl N, Marissen WE, Loeb M, Kelvin DJ, Preiser W, ter Meulen J, de Kruif J (2006) Isolation and characterization of human monoclonal antibodies from individuals infected with West Nile virus. J Virol 80(14):6982–6992

35. Vogt MR, Moesker B, Goudsmit J, Jongeneelen M, Austin SK, Oliphant T, Nelson S, Pierson TC, Wilschut J, Throsby M, Diamond MS (2009) Human monoclonal antibodies against West Nile virus induced by natural infection neutralize at a postattachment step. J Virol 83(13):6494–6507

36. Goo L, Debbink K, Kose N, Sapparapu G, Doyle MP, Wessel AW, Richner JM, Burgomaster KE, Larman BC, Dowd KA, Diamond MS, Crowe JE Jr, Pierson TC (2019) A protective human monoclonal antibody targeting the West Nile virus E protein preferentially recognizes mature virions. Nat Microbiol 4(1):71–77. https://doi.org/10.1038/s41564-018-0283-7

37. Narayan R, Tripathi S (2020) Intrinsic ADE: the dark side of antibody dependent enhancement during dengue infection. Front Cell Infect Microbiol 10:580096. https://doi.org/10.3389/fcimb.2020.580096

38. Byrne AB, Talarico LB (2021) Role of the complement system in antibody-dependent enhancement of flavivirus infections. Int J Infect Dis 103:404–411. https://doi.org/10.1016/j.ijid.2020.12.039

39. Bournazos S, Gupta A, Ravetch JV (2020) The role of IgG Fc receptors in antibody-dependent enhancement. Nat Rev Immunol 20(10):633–643. https://doi.org/10.1038/s41577-020-00410-0

40. Capell T, Twyman RM, Armario-Najera V, Ma JK, Schillberg S, Christou P (2020) Potential applications of plant biotechnology against SARS-CoV-2. Trends Plant Sci 25(7):635–643. https://doi.org/10.1016/j.tplants.2020.04.009

41. Chen Q (2008) Expression and purification of pharmaceutical proteins in plants. Biol Eng 1(4):291–321. https://doi.org/10.13031/2013.26854

42. Chen Q, Davis KR (2016) The potential of plants as a system for the development and production of human biologics. F1000Res 5. https://doi.org/10.12688/f1000research.8010.1

43. Lico C, Chen Q, Santi L (2008) Viral vectors for production of recombinant proteins in plants. J Cell Physiol 216(2):366–377

44. Chen Q (2016) Glycoengineering of plants yields glycoproteins with polysialylation and other defined N-glycoforms. Proc Natl Acad Sci U S A 113(34):9404–9406. https://doi.org/10.1073/pnas.1610803113

45. Chilton MD, Drummond MH, Merio DJ, Sciaky D, Montoya AL, Gordon MP, Nester EW (1977) Stable incorporation of plasmid DNA into higher plant cells: the molecular basis of crown gall tumorigenesis. Cell 11:263–271

46. Chen Q, He J, Phoolcharoen W, Mason HS (2011) Geminiviral vectors based on bean yellow dwarf virus for production of vaccine antigens and monoclonal antibodies in plants. Hum Vaccin 7(3):331–338. https://doi.org/10.4161/hv.7.3.14262

47. Lico C, Chen Q, Santi L (2008) Viral vectors for production of recombinant proteins in plants. J Cell Physiol 216(2):366–377. https://doi.org/10.1002/jcp.21423

48. Chen Q, Lai H, Hurtado J, Stahnke J, Leuzinger K, Dent M (2013) Agroinfiltration as an effective and scalable strategy of gene delivery for production of pharmaceutical proteins. Adv Tech Biol Med 1(1). https://doi.org/10.4172/atbm.1000103

49. Peyret H, Lomonossoff GP (2015) When plant virology met agrobacterium: the rise of the deconstructed clones. Plant Biotechnol J 13(8):1121–1135. https://doi.org/10.1111/pbi.12412

50. Klimyuk V, Pogue G, Herz S, Butler J, Haydon H (2014) Production of recombinant antigens and antibodies in Nicotiana benthamiana using 'magnifection' technology: GMP-compliant facilities for small- and large-scale manufacturing. Curr Top Microbiol Immunol 375:127–154. https://doi.org/10.1007/82_2012_212

51. Chen Q, He J, Phoolcharoen W, Mason HS (2011) Geminiviral vectors based on bean yellow dwarf virus for production of vaccine antigens and monoclonal antibodies in plants. Hum Vaccin 7(3):331–338

52. Diamos AG, Hunter JGL, Pardhe MD, Rosenthal SH, Sun H, Foster BC, DiPalma MP, Chen Q, Mason HS (2020) High level production of monoclonal antibodies using an optimized plant expression system. Front Bioeng Biotechnol 7(472). https://doi.org/10.3389/fbioe.2019.00472

53. Chen Q, Lai H (2015) Gene delivery into plant cells for recombinant protein production. Biomed Res Int 2015:932161. https://doi.org/10.1155/2015/932161

54. Chen Q (2022) Development of plant-made monoclonal antibodies against viral infections. Curr Opin Virol 52:148–160. https://doi.org/10.1016/j.coviro.2021.12.005

55. Jugler C, Sun H, Chen Q (2021) SARS-CoV-2 spike protein-induced interleukin 6 signaling is blocked by a plant-produced anti-interleukin 6 receptor monoclonal antibody. Vaccine 9(11):1365

56. Kallolimath S, Sun L, Palt R, Stiasny K, Mayrhofer P, Gruber C, Kogelmann B, Chen Q, Steinkellner H (2021) Highly active engineered IgG3 antibodies against SARS-CoV-2. Proc Natl Acad Sci 118(42):e2107249118. https://doi.org/10.1073/pnas.2107249118

57. He J, Lai H, Esqueda A, Chen Q (2021) Plant-produced antigen displaying virus-like particles evokes potent antibody responses against West

Nile virus in mice. Vaccine 9(1):60. https://doi.org/10.3390/vaccines9010060

58. Diamos AG, Pardhe M, Sun H, Mor TS, Chen Q, Mason H (2020) Codelivery of improved immune complex and virus-like particle vaccines containing Zika virus envelope domain III synergistically enhances immunogenicity. Vacccine 38(18):3455–3463

59. Hurtado J, Acharya D, Lai H, Sun H, Kallolimath S, Steinkellner H, Bai F, Chen Q (2020) In vitro and in vivo efficacy of anti-chikungunya virus monoclonal antibodies produced in wild-type and glycoengineered Nicotiana benthamiana plants. Plant Biotechnol J 18(1):266–273. https://doi.org/10.1111/pbi.13194

60. Lai H, Paul AM, Sun H, He J, Yang M, Bai F, Chen Q (2018) A plant-produced vaccine protects mice against lethal West Nile virus infection without enhancing Zika or dengue virus infectivity. Vaccine 36(14):1846–1852. https://doi.org/10.1016/j.vaccine.2018.02.073

61. Yang M, Sun H, Lai H, Hurtado J, Chen Q (2018) Plant-produced Zika virus envelope protein elicits neutralizing immune responses that correlate with protective immunity against Zika virus in mice. Plant Biotechnol J 16(2):572–580. https://doi.org/10.1111/pbi.12796

62. Lai H, Engle M, Fuchs A, Keller T, Johnson S, Gorlatov S, Diamond MS, Chen Q (2010) Monoclonal antibody produced in plants efficiently treats West Nile virus infection in mice. Proc Natl Acad Sci U S A 107(6):2419–2424. https://doi.org/10.1073/pnas.0914503107

63. Jugler C, Joensuu J, Chen Q (2020) Hydrophobin-protein a fusion protein produced in plants efficiently purified an anti-West Nile virus monoclonal antibody from plant extracts via aqueous two-phase separation. Int J Mol Sci 21(6):2140

64. Fulton A, Lai H, Chen Q, Zhang C (2015) Purification of monoclonal antibody against Ebola GP1 protein expressed in Nicotiana benthamiana. J Chromatogr A 1389:128–132. https://doi.org/10.1016/j.chroma.2015.02.013

65. Lai H, He J, Engle M, Diamond MS, Chen Q (2012) Robust production of virus-like particles and monoclonal antibodies with geminiviral replicon vectors in lettuce. Plant Biotechnol J 10(1):95–104. https://doi.org/10.1111/j.1467-7652.2011.00649.x

66. Chen Q, Dent M, Hurtado J, Stahnke J, McNulty A, Leuzinger K, Lai H (2016) Transient protein expression by agroinfiltration in

lettuce. Methods Mol Biol 1385:55–67. https://doi.org/10.1007/978-1-4939-3289-4_4

67. Strasser R, Stadlmann J, Schahs M, Stiegler G, Quendler H, Mach L, Glossl J, Weterings K, Pabst M, Steinkellner H (2008) Generation of glyco-engineered Nicotiana benthamiana for the production of monoclonal antibodies with a homogeneous human-like N-glycan structure. Plant Biotechnol J 6(4):392–402. https://doi.org/10.1111/j.1467-7652.2008.00330.x

68. Chen Q (2016) Glycoengineering of plants yields glycoproteins with polysialylation and other defined N-glycoforms. Proc Natl Acad Sci 113(34):9404–9406. https://doi.org/10.1073/pnas.1610803113

69. Lai H, He J, Hurtado J, Stahnke J, Fuchs A, Mehlhop E, Gorlatov S, Loos A, Diamond MS, Chen Q (2014) Structural and functional characterization of an anti-West Nile virus monoclonal antibody and its single-chain variant produced in glycoengineered plants. Plant Biotechnol J 12(8):1098–1107. https://doi.org/10.1111/pbi.12217

70. Watts RJ, Dennis MS (2013) Bispecific antibodies for delivery into the brain. Curr Opin Chem Biol 17(3):393–399. https://doi.org/10.1016/j.cbpa.2013.03.023

71. Esqueda A, Jugler C, Chen Q (2021) Design and expression of a bispecific antibody against dengue and chikungunya virus in plants. Methods Enzymol. https://doi.org/10.1016/bs.mie.2021.05.003

72. He J, Lai H, Engle M, Gorlatov S, Gruber C, Steinkellner H, Diamond MS, Chen Q (2014) Generation and analysis of novel plant-derived antibody-based therapeutic molecules against West Nile virus. PLoS One 9(3):e93541. https://doi.org/10.1371/journal.pone.0093541

73. Dent M, Hurtado J, Paul AM, Sun H, Lai H, Yang M, Esqueda A, Bai F, Steinkellner H, Chen Q (2016) Plant-produced anti-dengue virus monoclonal antibodies exhibit reduced antibody-dependent enhancement of infection activity. J Gen Virol 97(12):3280–3290. https://doi.org/10.1099/jgv.0.000635

74. Group TPIW (2016) A randomized, controlled trial of ZMapp for Ebola virus infection. N Engl J Med 375(15):1448–1456. https://doi.org/10.1056/NEJMoa1604330

75. Ma JKC, Drossard J, Lewis D, Altmann F, Boyle J, Christou P, Cole T, Dale P, van Dolleweerd CJ, Isitt V, Katinger D, Lobedan M, Mertens H, Paul MJ, Rademacher T, Sack M, Hundleby PAC, Stiegler G, Stoger E, Twyman RM, Vcelar B, Fischer R (2015) Regulatory approval and a first-in-human phase I clinical trial of a monoclonal antibody produced in transgenic tobacco plants. Plant Biotechnol J 13(8):1106–1120. https://doi.org/10.1111/pbi.12416

76. Traynor K (2012) Taliglucerase alfa approved for Gaucher disease. Am J Health Syst Pharm 69(12):1009. https://doi.org/10.2146/news120041

INDEX

Fengwei Bai (ed.), *West Nile Virus: Methods and Protocols*,
Methods in Molecular Biology, vol. 2585, https://doi.org/10.1007/978-1-0716-2760-0,
© The Editor(s) (if applicable) and The Author(s), under exclusive license to Springer Science+Business Media, LLC, part of Springer
Nature 2023

Milton Keynes UK
Ingram Content Group UK Ltd.
UKHW050646101123
432316UK00004B/18